ADVANCES IN

Combustion Toxicology

VOLUME ONE

Edited by Gordon E. Hartzell

CRC Press
Taylor & Francis Group
Boca Raton London New York

CRC Press is an imprint of the
Taylor & Francis Group, an **informa** business

First published 1989 by Technomic Publishing Company, Inc.

Published 2019 by CRC Press
Taylor & Francis Group
6000 Broken Sound Parkway NW, Suite 300
Boca Raton, FL 33487-2742

© 1989 by Taylor & Francis Group, LLC
CRC Press is an imprint of Taylor & Francis Group, an Informa business

First issued in paperback 2019

No claim to original U.S. Government works

ISBN-13: 978-0-367-45122-6 (pbk)
ISBN-13: 978-0-87762-590-2 (hbk)

Main entry under title:
    Advances in Combustion Toxicology—Volume One

A Technomic Publishing Company book
Bibliography: p.
Includes index p. 323

ISSN No. 0899-7195

Table of Contents

Preface

NUMEROUS HAZARDS TO humans result from exposure to the products of combustion of fires. Predominant among these are effects from heat and flames, visual obscuration due to the density of smoke or to eye irritation, narcosis from inhalation of asphyxiants and irritation of the upper and/or lower respiratory tracts. These effects, often occurring almost simultaneously in a fire, contribute to physical incapacitation, loss of motor coordination, faulty judgement, disorientation, restricted vision, and panic. The resulting delay or prevention of escape may lead to subsequent injury or death from further inhalation of toxic gases and/or the suffering of thermal burns. Survivors from a fire situation may also experience post-exposure pulmonary complications and burn injuries which can lead to delayed death.

It is widely acknowledged that, in most fires, obscuration of vision by smoke is the first threat to safe egress. However, visual obscuration *per se* is not necessarily a dangerous physiological threat to life. Furthermore, the hazards of thermal injury or burns are usually present only in the proximity of the fire itself. It is the physiological and behavioral effects of the inhalation of smoke which can be manifest even at considerable distances from the fire and which have created the most attention and concern among the fire safety experts and the general public.

Assessment of the overall physiological and behavioral effects of exposure of humans to fire and its combustion products is an extremely difficult and complex task—a task which has often involved considerable speculation and a variety of opinions, even among the experts. However, research in combustion toxicology over the past few years has led to a reasonable understanding and even quantification of some of the effects of fire effluent toxicants and, with the availability of a modest amount of both non-human primate and human exposure data, the combustion toxicologist is gaining increasing capability to assess and predict the toxicological effects of smoke inhalation. Furthermore, the roles of individual materials and also specific fire scenarios as they relate to human hazard are becoming better understood.

It is the intent of *Advances in Combustion Toxicology* to periodically assemble technical papers on the subject of combustion toxicology as originally published in the *Journal of Fire Sciences* and to provide up-to-date editorial commentary on the state-of-the-art at the time of publication of each volume. The commentary will address the three basic areas of combustion toxicology:

- Combustion of Materials
- Assessment of the Toxicity of Smoke
- Understanding of Hazards to Humans

The first two volumes of the series include papers published in the *Journal of Fire Sciences* during 1983–1987. Technical papers published prior to 1983 were reviewed and/or referenced in *Combustion Toxicology: Principles and Test Methods*, written by H. L. Kaplan, A. F. Grand and G. E. Hartzell, and published in 1983 by Technomic Publishing Company, Incorporated. Thus, between *Combustion Toxicology: Principles and Test Methods* and *Advances in Combustion Toxicology*, one may have ready access to most of the published research and test data in the field of combustion toxicology. Coverage of subject matter includes contributions from many of the leading authorities in the field from the United States, Europe, Canada and Japan.

About the Editor

DR. GORDON E. HARTZELL has been director of the Department of Fire Technology at SwRI since 1978. His experience includes 20 years of industrial research and development in organic and polymer chemistry at the Dow Chemical Company. Before joining SwRI, he was a Research Associate Professor of Materials Science at the University of Utah.

Active in the field of fire technology since 1974, Dr. Hartzell has authored over 50 papers, largely in the areas of combustion toxicology and hazards to life due to fire. He is a member of American Society for Testing and Materials Committee E-5 on Fire Standards and serves on working groups of International Organization for Standardization (ISO) TC92/SC3 which address toxic hazards in fire.

Dr. Hartzell is a co-author of the book, *Combustion Toxicology: Principles and Test Methods*, and is Editor-in-Chief of the *Journal of Fire Sciences*.

Combustion of Materials

G. E. HARTZELL

Department of Fire Technology
Southwest Research Institute
P.O. Drawer 28510
San Antonio, Texas 78284

FIRE IS A complex and interrelated array of physical and chemical phenomena. As a result, it is extremely difficult to simulate a real fire on a laboratory scale suitable for controlled test procedures. This is perhaps the single most perplexing problem associated with combustion toxicology: the combustion process—the way it is done experimentally in the laboratory relative to the actual combustion of materials in a "real" fire. The problem is common to all of fire testing, but it is in combustion toxicology that additional restrictions and criteria are imposed due to the need for the laboratory combustion to be compatable with bioassay procedures using live test animals. For example, the reduced oxygen levels involved and heat produced must not, in themselves, be unduly compromising to exposed animals. At the same time, sufficiently high concentrations of toxicants must be produced so as to obtain toxicological effects within a reasonable exposure time. As a result of these restrictions, many compromises must be made in the combustion system which may further reduce the apparent validity of any combustion methodology. Many object to testing for the toxicity of smoke on the grounds that no laboratory test can relate to the performance of materials in a real fire.

It is beyond the scope of this commentary to go into detailed analysis of the physics and chemistry of fire. J. W. Lyons presents a particularly broad and well-written discussion of the subject in *Fire*, published in 1985 [1]. Relevant to combustion toxicology, the reader is referred to the paper by Huggett [2] and also to the recent International Organization for Standardization (ISO) Technical Report 9122, "Toxicity Testing of Fire Effluents: The State-of-the-Art 1987." *Combustion Toxicology:*

Principles and Test Methods, by Kaplan, Grand and Hartzell, reviewed in detail the various laboratory combustion devices used for smoke toxicity testing [3].

For the purposes of this discussion on combustion toxicology, the following combustion modes are generally felt to encompass the relevant range of fire characteristics:

Flaming
> Fuel Controlled—adequate oxygen supply
> Ventilation Controlled
>> Ventilation limited
>> Freely ventilated

Nonflaming
> Oxidative—thermal decomposition in the presence of oxygen
> Pyrolytic—thermal decomposition in the absence of oxygen

Smoldering—self-propagating thermal decomposition

The primary chemical process leading to formation of combustion products is that of the thermal bond-breaking and decomposition of polymeric materials which, in the presence of oxygen, may lead to a variety of oxygenated species. Depending on the oxygen supply, carbon may be oxidized to carbon monoxide or carbon dioxide. Both are usually present in a fire effluent atmosphere, with the ratio of the two often being used as an indicator characteristic of the particular type or stage of a fire. For example, a ratio of CO_2/CO of 20 or more in a fully developed, ventilation-controlled fire would indicate freely ventilated combustion, whereas a ratio of 10 to 20 would suggest limited ventilation. Hydrogen is oxidized to water; chlorine is most commonly released as hydrogen chloride; and nitrogen appears as nitrogenous organic compounds, hydrogen cyanide and/or nitrogen oxides, depending upon the thermal and oxidative conditions. Nonflaming and smoldering fires may yield a myriad of combustion products due to incomplete decomposition and only partial oxidation of the fuels involved. Important to remember is that these are all chemical reactions, subject to the usual principles of thermodynamics and kinetics. Thus, stoichiometry and thermal energy play significant roles in determining the products of combustion that are formed during the various stages of a fire and over the range of fire classes.

In an attempt to simulate as many of the fire conditions as possible, combustion toxicologists have used a variety of laboratory devices, including cups, crucibles, tubes and cones of one sort or another in which test specimens are subjected to radiant, conductive and/or convective heat. All too often, an investigator has simply utilized some device which was available in the laboratory, with insufficient attention being paid to the physics and dynamics of fire. Thus far, only the moving annular tube furnace of the DIN (German Standards Institution) 53436 [4] and the radiant furnace modification [5] of the NBS (National Bureau of Standards) test method seem to have been designed especially for use in combustion toxicology. Other combustion devices, usually known by the smoke toxicity test with which they are associated, have also achieved a significant level of attention. These are the furnaces used in the NBS (National Bureau of Standards) [6], UPITT (University of Pittsburgh) [7] and JPN-2 BRI (Japan-Building Research Institute) and RIPT (Japan-Research Institute for Polymers and Textiles) tests [8]. The following will serve to describe briefly the common combustion devices. (The DIN 53456 and the US-radiant furnace are also described in detail in papers contained in this *Advances in Combustion Toxicology* series.)

The NBS test [6] employs a cup or crucible furnace, often referred to as the "Potts furnace," named after the investigator who first reported its use in combustion toxicology. Heating is considered to be largely conductive, with the bottom and lower portion of the quartz cup constituting the hot zone. Test materials of up to 8 grams are introduced into the cup which has a volume of about one liter. Procedures for testing materials involve combustion at both just below (nonflaming) and just above (flaming) an autoignition temperature. There has been particular concern regarding air flow into the cup and, although it does communicate with a volume of 200 L of air contained in the exposure chamber, it is uncertain as to the adequacy of the oxygen supply for combustion. With the variable sample sizes used and an ill-defined fuel/air ratio, some feel the system fails to carry out combustion in a well-characterized manner. A further criticism has been that various materials are not tested under the same conditions but often at considerably different temperatures.

Radiant heating is a major source of energy transfer in a real fire; therefore, a radiant furnace combustion device may represent one of the best laboratory-scale combustion models for toxicology studies. Furthermore, in the case of composite materials, exposure could be conducted upon the normally exposed surface in much the same manner as in a real fire. Such a radiant furnace device has been used as an alternative to the NBS cup furnace for combustion toxicology studies [5]. Specimens may undergo either flaming or nonflaming combustion, depending on the imposed heat flux and the presence of an ignition source. In this device, the combustion cell's relative dimensions are such that quite effective mixing takes place with the reservoir of air in the animal exposure chamber, especially under flaming conditions. A major disadvantage in the system is the laboratory time required to conduct tests due, in part, to the requirement that the combustion cell be disassembled and reassembled after each test and also to the frequent heat flux calibrations required. There is also some problem with the furnace windows becoming clouded by smoke during testing.

The DIN 53436 combustion device [4] is characterized by the use of a moving annular tube furnace operated at a constant temperature from 200 to 600°C, with fire effluents being diluted with air as a means of varying concentration. Differing from some real fires, the air flow is counter to the direction of flame propagation. The DIN furnace offers a rather wide range of well-controlled combustion conditions and is particularly suited to simulate the nonflaming fire condition. Some difficulty has been experienced in controlling smoldering and flaming conditions; however, recent work using sample segmentation has shown promise with these modes of combustion [9]. Neither the DIN nor the NBS combustion devices provide for the continuous monitoring of sample weight. This is a distinct disadvantage if it is desired to calculate a summation of incremental exposure doses to which test animals are subjected.

In the UPITT test method [7], the combustion device is a muffle or box furnace which is often used in an inverted position in order to provide for a pedestal connected to a mass sensor. With this arrangement, continuous monitoring of sample weight is conducted. Combustion is accomplished using a linearly ramped temperature regime of 20°C per minute up to as high as 1100°C, while 11 L/min of air is pulled through the furnace. Smoke concentration is varied by changing the weight of material charged to the furnace. The fuel/air ratio can, therefore, vary widely depending on the weight of sample and also on its rate of combustion. Huggett has suggested that after a sample ignites the combustion device may simulate conditions encountered in real fires better than some of the other combustion schemes [2]. A problem of explosions has been encountered with some materials which thermally decompose or burn very rapidly once ignition occurs. Such explosions, possibly the rapid emission of gases whose volume exceeds that being pulled from the furnace, have not been reported to be sufficiently severe as to rupture the apparatus; however, proper caution should be exercised.

The most recently reported combustion method used in Japan for smoke toxicity testing is that associated with the JPN-2 BRI (Building Research Institute) test [8]. The combustion device consists of a cone-shaped quartz tube positioned inside a cone radiative electric furnace. External radiant heat fluxes up to 4.5 W/cm^2 may be used, with combustion normally being in the flaming mode. An electric ignitor is used to initiate the flaming as early as possible. Air, supplied through the bottom of the quartz tube, flows around the sample during combustion. The air supply is varied so as to sustain flaming at a minimum rate. The mass loss of sample is monitored continuously with a load cell. There has been little experience outside of Japan using this combustion device, but the methodology would appear to be reasonable for its purpose.

Another combustion device under development in Japan is a cone-type radiant furnace used in a test known as the JPN-2 RIPT (Research Institute for Polymers and Textiles) [8]. Details are rather similar to those for the BRI method except for the capability to vary the oxygen concentration in the combustion chamber.

The best approach to the selection of an appropriate combustion device for smoke toxicity testing involves the consideration of data which would relate laboratory combustion conditions to the types of stages of real fires. Recognizing that no laboratory device can likely be expected to relate to all types of fires, there has been strong sentiment for a prioritization favoring the well-developed ventilation-controlled fire as being of primary importance. The validity of this choice is evident from numerous full-scale fire tests in which the toxic threat is not developed until flashover occurs [10]. (Although the smoldering fire can present a definite toxic hazard given sufficient time, it cannot be satisfactorily modeled in the laboratory.) With criteria limited to the conditions for either a limited ventilation or a freely ventilated, well-developed fire, a number of the laboratory combustion models may be quite satisfactory. This is currently under examination within the Toxic Hazards in Fire Subcommittee in ISO.

All of the laboratory combustion methods have critics as well as supporters. Since the laboratory combustion devices are primarily used for the testing of materials, whether or not a combustion device is a good one may often be determined largely "in the eye of the beholder." That is to say, if a producer's product tests well, it's a good test; if not, it's a poor test. Surprisingly, toxicity test results, if expressed in comparable units of exposure dosage required to obtain a lethal endpoint, would appear to be rather insensitive to the combustion device and protocol used. This may be due to most laboratory combustion devices possibly modeling similar types of fires or a matter of the variability of bioassay methodology being sufficiently great as to obscure much of the differences due to combustion methods.

It is the opinion of this author that there are overriding considerations of tradition as to the "suitability" of a particular combustion device. The DIN 53436 furnace is firmly entrenched in Europe and will be unlikely to be replaced with any U.S. device. Similarly, the UPITT-box furnace is well established in the requirements of the state of New York [11] and, as more and more test data are filed there, it will be increasingly difficult to displace its use in the New York regulations. Furthermore, operators of the UPITT test appear to like the method due to its simplicity and reliability. The NBS cup furnace method does not have a champion, not even the Bureau of Standards itself. The NBS test procedure, furthermore, is more than twice as expensive to conduct as the UPITT method.

A few studies have been conducted using full-scale fires to evaluate the contribution of certain construction materials and furnishings to toxic hazard [10,12,13]. Considerable caution should be exercised in drawing conclusions from these studies. The development of toxic hazard depends on fire growth which is very difficult to replicate on a large scale. It can be done under certain circumstances, but often large amounts of toxicity-related data are obtained which are meaningless for comparative purposes simply because the fires were different. Furthermore, such large-scale tests are quite expensive and researchers usually cannot afford to do sufficient work to draw fully valid conclusions.

REFERENCES

1. Lyons, J. W. *Fire*, New York, N.Y.:Scientific American Books, Inc., p. 7 (1985).
2. Huggett, C. "Combustion Conditions and Exposure Conditions for Combustion Product Toxicity Testing," *J. Fire Sciences*, 2(5):328–347 (September/October 1984).
3. Kaplan, H. L., A. F. Grand and G. E. Hartzell. *Combustion Toxicology: Principles and Test Methods*, Lancaster, PA:Technomic Publishing Company, Inc. (1983).
4. German Standards Institution. DIN 53436, "Erzeugung thermischer Zersetzungsproduckte von Werkstoffen unter Luftzufuhr und ihre toxikologische Prufung. Part 1, Zersetzungsgerat und Bestimmung der Versuchstemperatur. Part 2, Verfahren zur thermischen Zersetzung" (January 1979).
5. Alexeeff, G. V. and S. C. Packham. "Use of a Radiant Furnace Fire Model to Evaluate Acute Toxicity of Smoke," *J. Fire Sciences*, 2(4):306–320 (July/August 1984).
6. Levin, B. C., A. J. Fowell, M. M. Birky, M. Paabo, A. Stolte and D. Malek. "Further Development of a Test Method for the Assessment of the Acute Inhalation Toxicity of Combustion Products," NBSIR 82-2532, National Bureau of Standards, Gaithersburg, MD (1982).
7. Kaplan, H. L., A. F. Grand and G. E. Hartzell. *Combustion Toxicology: Principles and Test Methods*, Lancaster, PA:Technomic Publishing Company,

Inc., pp. 86–87 (1983); Alarie, Y. C. and R. C. Anderson. "Toxicologic and Acute Lethal Hazard Evaluation of Thermal Decomposition Products of Synthetic and Natural Polymers," *Toxicology and Applied Pharmacology*, 51:341–362 (August 1979).

8. Saito, F. and S. Yusa. "Toxicity Testing of Fire Effluents in Japan: State-of-the-Art Review," *Fire Safety Science—Proceedings of the First International Symposium*, New York, N.Y.:Hemisphere Publishing Corporation, pp. 1069–1077 (1986).

9. Prager, F. H., H. J. Einbrodt, J. Hupfeld, B. Muller and H. Sand. "Risk Oriented Evaluation of Fire Gas Toxicity Based on Laboratory Scale Experiments—The DIN 53436 Method," *J. Fire Sciences*, 5(5):308–325 (September/October 1987).

10. Grand, A. F., H. L. Kaplan, J. J. Beitel, III, W. G. Switzer and G. E. Hartzell. "An Evaluation of Toxic Hazards from Full-Scale Furnished Room Fire Studies," *Fire Safety: Science and Engineering*, ASTM STP 882, T. Z. Harmathy, ed. Philadelphia, PA:American Society for Testing and Materials, pp. 330–353 (1985).

11. "Combustion Toxicity Testing," New York State Uniform Fire Prevention and Building Code, Article 15, Part 1120, Office of Fire Prevention and Control, Department of State, New York State, Albany, NY (1986).

12. Morikawa, T. and E. Yanai. "Toxic Gases Evolution from Air-Controlled Fires in a Semi-Full Scale Room," *J. Fire Sciences*, 4(5):299–314 (September/October 1986).

13. Morikawa, T., E. Yanai and T. Nishina. "Toxicity Evaluation of Fire Effluent Gases from Experimental Fires in a Building," *J. Fire Sciences*, 5(4):248–271 (1987).

2

Assessment of the Toxicity of Smoke

G. E. HARTZELL

Department of Fire Technology
Southwest Research Institute
P.O. Drawer 28510
San Antonio, Texas 78284

ALTHOUGH CONCERN REGARDING the incidence of fire fatalities from inhalation of toxic smoke had been expressed in the literature as early as the 1930s [1], it remained for Zapp to pioneer the field of fire toxicology with extensive medical-physiological investigations reported in 1951 [2]. A number of studies were conducted over the following years, but it was not until the 1970s that more serious attention was directed toward combustion toxicology, with reviews by Kimmerle in 1973 [3] and Birky in 1976 and 1977 [4,5]. A report on "Fire Toxicology: Methods for Evaluation of Toxicity of Pyrolysis and Combustion Products" by a National Academy of Sciences Committee on Fire Toxicology was issued in 1977 [6]. The field was again reviewed by Kaplan, Grand and Hartzell in 1983 in *Combustion Toxicology: Principles and Test Methods* [7]. In 1986, another National Academy of Sciences Committee on Fire Toxicology published a further report on "Fire and Smoke: Understanding the Hazards [8]."

Advances in assessment of the toxicity of smoke have basically progressed along two somewhat parallel courses—the development of methods for the testing of materials and the development of models for predicting the toxicity of smoke from analytical data obtained in studies involving single or multiple fire gas toxicants.

MATERIAL TESTING

The serious development of smoke toxicity test methods for materials began in the U.S. in the mid-1970s, primarily under the impetus of a

Federal Trade Commission consent order involving the cellular plastics industry [9]. The two methods most widely used in the U.S., usually referred to as the NBS (National Bureau of Standards) and the UPITT (University of Pittsburgh) methods, had their origins under the FTC consent order. The NBS test method stemmed from research conducted by a number of workers at the University of Utah, whereas Professor Y. C. Alarie and his students developed the UPITT methodology at the University of Pittsburgh. At approximately the same time, test methods were also being developed in Japan and West Germany.

Most methods for assessment of the smoke toxicity of materials are bioassay (animal exposure) methods, supplemented with chemical analyses. However, methods using only analyses are being employed for the selection of materials by some government agencies and manufacturing industries. One of these which has received considerable attention is the British Naval Engineering Standard 713 [10].

In the bioassays, the toxicity of smoke is evaluated on the basis of measurement of the response of test animals when exposed to the smoke for a specified period of time. Rodents, usually rats or mice, are normally used. Lethality is the most commonly measured animal response, although some test methods obtain a measure of incapacitation of the animal. For any biological response, the relationship is determined between the response of the test animals and different concentrations of the smoke. This is accomplished by conducting a series of experiments in which the quantity of material combusted or the flow rate of diluting air is varied in order to produce different concentrations of smoke. The number of animals showing a response, either as lethality or incapacitation, will increase as the concentration is increased. In combustion toxicology, concentration is traditionally expressed either as the quantity of test material used per chamber unit volume (the material charge concentration) or the material weight loss per chamber unit volume (the smoke concentration). Test methods which employ dynamic or flow-through generation of smoke usually express concentration simply as the weight of material charged. (The weight per unit volume can be calculated from the air flow rate.) When the percent of animals responding within a specified time is plotted as a function of the logarithm of the concentration, a straight line is approximated. Such a plot represents a concentration-response relationship of the smoke produced by a material under the experimental conditions of the particular test method. The concentration which would produce a response or effect in 50 percent of the animals upon exposure for the specified time is obtained from the data using a statistical calculation. This concentration, commonly termed the EC_{50}, is a measure of the potency of the smoke. The EC_{50} is a general term and may be used in reference to any measured response of the animal. When lethality is the measured response, the LC_{50} is used as a more specific term to denote the concentration of material or smoke that produces death in 50 percent of the animals. The LC_{50} is the most commonly reported measurement of toxic potency in smoke toxicity testing, although some test methods do report other response measurements.

A brief description of the major methods actually in significant use is as follows:

The DIN 53436 method (Federal Republic of Germany-Deutsches Institut fuer Normung) specifies the combustion apparatus and the operating procedure [11]. The description of the animal response model is designated as a draft. The method is characterized by the use of a moving annular tube furnace operated at a constant temperature with fire effluents being diluted with air. Fire effluents are generated dynamically, i.e., continuously over the exposure time of the animals. Concentration-response relationships are easily obtainable, with concentration being varied by dilution of the fire effluent before the rats are exposed. The test is used primarily under nonflaming conditions, with rat lethality as the bioassay.

The NBS test (U.S.-National Bureau of Standards) is a static (closed) system using a cup-type furnace as the combustion device, operated just below and just above the auto-ignition temperature of the specimen [12]. Concentration-response (lethality) relationships for 30-minute exposures (14-day postexposure observation) are determined using rats held in tubular restrainers for head-only exposure. Six rats are used per test. Concentration is controlled by varying the sample weight charged to the cup. Data for both nonflaming and flaming combustion are obtained.

The U.S.-Rad (Radiant Furnace Test) uses the same exposure chamber as the NBS test, providing a static (closed) animal exposure system [13]. Animal exposures are controlled by varying the irradiation intensity and duration and are quantified in terms of concentration × time products. Rats are used after preliminary tests evaluating the relative concentrations of known toxicants are conducted.

The UPITT methodology (U.S.-University of Pittsburgh) exposes four mice per test to fire effluents produced in a dynamic (flow-through) system from the 20°C per minute ramped heating of a specimen in a box or muffle furnace [14,15]. Exposures are for 30 minutes (plus 10 minutes postexposure), starting when 1 percent of the sample weight has been lost. Bioassays include concentration-response and time-to-death for lethality and respiratory rate depression for assessment of irritants. In the UPITT test, the concentration term in the LC_{50} refers only to the weight of test specimen charged to the furnace. Thus, LC_{50} values for this test are not directly comparable to those from the NBS method.

The JGBR (Japanese Government Building Regulation Toxicity Test (1976) employs a radiant heat furnace (modified UK BS 476 Part 6–Fire Propagation Test) and exposes mice to smoke produced in a dynamic system from the ramped heating of a specimen [16]. Incapacitation times of eight mice placed in rotary cages are determined and compared with those of red lauan wood as a reference material. Two new types of test apparatus using radiant heat cone-type heaters are also being used in research studies in Japan. Both expose mice in rotary cages for determination of the time to incapacitation.

The DIN, UPITT and JGBR tests have regulatory application in the Federal Republic of Germany, the U.S. (state of New York only) [15] and Japan, respectively. Both the UPITT and the NBS tests are also used for regulatory purposes in the city of New York. It is beyond the scope of this review to examine the details of the regulatory use of these test methods.

All the test methods have been subjected to the following criticisms, particularly when used for regulatory purposes.

1. The relevance of laboratory combustion methodology to the conditions present in any real fire has not been adequately shown.
2. The applicability of laboratory animal responses to anticipate the conditions of humans being severely compromised in a fire exposure has not been demonstrated sufficiently.
3. The use of live animals for test purposes is opposed (and even prohibited) in many areas of the world.
4. Most materials do not exhibit significant differences in smoke toxicity from the norm.
5. The toxic hazard in a real fire depends more on fire growth than on the smoke toxicities of the individual materials being combusted.

In spite of these criticisms and limitations, laboratory smoke toxicity tests are becoming used for regulatory as well as for product research and development purposes. Experience thus far in the U.S. leads this writer to feel that implementation of requirements for the testing of products for toxicity of smoke is being done more as a matter of principle, rather than as one of gaining or denying approval of products. That is, the key objective is that products should be tested and the results only be reported. Proponents of such testing see this as a way of directing attention to the development of safer products from the point of view of smoke toxicity. It is hoped that other fire properties are not sacrificed along the way.

As much of industry has feared, the product marketing use of toxicity test data has already been seen. There have been some implications that one product may be more or less safe than another, based on numerical differences in toxic potency values. This practice is not only unwise, but is statistically and scientifically not valid in many cases. One analysis of data generated using the UPITT and the NBS combustion toxicity protocols has claimed that 30 percent of the LC_{50} values from the UPITT test were not statistically valid, while with the NBS test, 26 percent and 50 percent of the LC_{50} values under flaming and nonflaming conditions, respectively, were not statistically valid [17]. Comparisons of the two test methods also showed that there was not agreement on ranking across protocols except for the very few with extreme toxicities. These results indicated that caution should be exercised in the review and interpretation of combustion toxicity data and that the tests are probably only useful in identifying consistently supertoxic materials. (This use of the tests was, in fact, the original objective in their development in the 1970s.)

It is widely recognized that toxic hazard is ultimately the real concern and that, in a real fire, the hazard depends more on fire growth than on the smoke toxicities of the individual materials involved. Current thinking is evolving which would combine toxic potency data with data on mass loss rates and times to ignition in an effort to devise methodology for assessment of the contribution of a material to toxic hazard [18,19]. These concepts, still under development, appear to show considerable promise in addressing a measure of relative hazard from readily obtainable laboratory data.

MODELING OF TOXICOLOGICAL EFFECTS

The second major thrust in the area of assessment of the toxicity of smoke has been in the development of mathematical models for predicting the toxicity of smoke from appropriate data on the composition of fire gases. The objectives of these efforts are twofold. Assessment of smoke toxicity from analytical data could obviate much of the use of live animals in conventional bioassay methodology. Furthermore, providing that both qualitative and quantitative differences in toxicological effects between laboratory animals and man are understood, such modeling methodology could also be used for estimating the time to development of a toxic hazard in either real or simulated fire scenarios.

The development of smoke toxicity modeling methodology began in the early 1970s, with concepts proposed by Y. Tsuchiya and K. Sumi at the National Research Council Laboratories in Canada [20] and by G. W. Armstrong at Southwest Research Institute [21]. A publication in 1981 by S. C. Packham and G. E. Hartzell on "Fundamentals of Combustion Toxicology in Fire Hazard Assessment" established a foundation for such modeling in the U.S. [22].

Basic to all the modeling techniques is some expression of the concentration of a toxicant relative to that concentration known to cause a particular toxic effect resulting from a given time of exposure. Lacking in some of the early development efforts was a clear concept of the "dose" of a toxicant, along with appreciation of its utility as a tool in modeling. Also lacking was a good base of toxicological data appropriate for short exposures to relatively high concentrations of toxicants.

Quantification of "dose" has been fundamental to the development of methodology for modeling the toxicological effects of inhalation of fire gases, whether it be by laboratory animals or by humans. Physiological responses are usually "dose-related," i.e., the magnitude of the effect increases with increasing dose or accumulated body burden of a physiologically active agent. Since the real dose of toxicants from smoke inhalation cannot be directly measured, the assumption is made that the dose is a function of smoke (or toxicant) concentration and exposure time [23]. This "dose" is really an expression of the insult to which a subject *is exposed*. The term "exposure dose" is probably more accurate and is becoming the preferred term in combustion toxicology.

Concentrations of common fire gas toxicants, such as carbon monoxide and hydrogen cyanide, are usually expressed as parts per million (ppm) by volume. Therefore, the exposure dose can be expressed as the product of the concentration and time, i.e., ppm-min. In the case of a changing concentration of a gaseous toxicant, the exposure dose is actually the integrated area under a concentration *vs.* time curve.

Often, the concentrations of fire gas toxicants may not be known. In that event, one can still deal with the concept of exposure dose as it applies to smoke. Since smoke concentration cannot be quantified, an approximation is made that the smoke concentration is proportional to the mass loss during a fire. The integrated area under a mass loss per unit volume *vs.* time curve thus becomes a measure of smoke exposure dose, e.g., mg-min/liter, lb-min/ft³, etc. [24,25]. Smoke exposure dose at any point in time can be calculated from data obtained from a laboratory combustion device, instrumented experimental fires, data generated from mathematically modeled fires and even data estimated from real fires. In the case of real fire scenarios, smoke transport models and dilution calculations can provide for estimation of smoke exposure doses even in areas remote from a fire. It is an important concept that "toxicological exposure doses" can be visualized as quantified entities that are generated in a fire, transported and then administered to exposed subjects.

The next step in the modeling of toxicological effects is to determine just what constitutes the exposure dose associated with a given response, e.g., lethality, incapacitation, etc. Furthermore, it is usually necessary to determine the dependence of that effective exposure dose on the concentration of a toxicant, since Haber's Rule (concentration × time = constant) may not be valid over the range of concentrations of interest. This is done from a concentration-time-response data base for the toxicant being studied. It has been found that, in general, the exposure dose required to cause a particular response decreases with increasing concentration of a toxicant. This is particularly the case with hydrogen cyanide (HCN) and hydrogen chloride (HCl). Haber's Rule appears to be most applicable to carbon monoxide (CO), with concentration × time being relatively constant over the range of interest in combustion toxicology.

Once effective exposure doses are characterized, extension of the concept to that of a "fractional effective dose" (FED), along with the summation or integration of FEDs, results in workable tools in combustion toxicology [26,27]. From toxicant concentration as a function of time data, incremental exposure doses (C × Δt) are calculated and related to the specific Ct exposure dose required to produce the given toxicological effect at each particular incremental concentration. Thus a "fractional effective dose" (FED) is calculated for each small time interval. Continuous summation of these "fractional effective doses" is carried out and the time at which this sum becomes unity (100 percent) represents the time at which 50 percent of the exposed subjects would be expected to incur the toxicological effect. The basic concepts of summing "fractional effective doses" are now practiced in a variety of forms by several research groups.

Most laboratory studies have involved the most prevalent gaseous fire effluents, i.e., CO, CO_2, O_2, HCN and HCl, with exposure doses associated with both lethality and incapacitation of rodents now being fairly well characterized. Much of the recent data is contained in publications by B. C. Levin [28] and by G. E. Hartzell [26,27,29,30,31,32]. Reference should be made to the original literature for these detailed data in order to fully understand the animal exposure methods used and, therefore, to assess the scope and limitations of the data.

Of greatest interest in the modeling of toxicological effects is how to accommodate atmospheres containing multiple toxicants. It now appears that, even though different toxicants may have quite different physiological mechanisms, fractional doses can often be additive in an empirical sense.

Although there is no evidence for synergistic effects between carbon monoxide and hydrogen cyanide [33], these two asphyxiant toxicants do appear to be additive when expressed as fractional doses required to cause either incapacitation or lethality [27,28]. Thus, as a reasonable approximation, the fraction of an effective dose of CO can be added to that of HCN and the time at which the sum becomes unity (100 percent) can be used to estimate the occurrence of a particular toxicological effect.

In the case of mixtures of hydrogen chloride and carbon monoxide, empirical analysis of toxicological data shows that exposure doses leading to lethality may also be additive [32]. Observations with rodents suggest that the animals are compromised both by respiratory acidosis (caused by HCl) and by metabolic acidosis (caused by CO). Postexposure recovery is usually slow, with COHb unloading being delayed and blood pH remaining low longer than normal. The result of the combined insult is an increased incidence of early postexposure deaths of the rats.

Incapacitation due to CO is an entirely different situation, however, in the presence of HCl. Even modest concentrations of HCl exhibit a sensory irritation effect, resulting in decreased respiratory minute volume (RMV) and a resulting decrease in the rate of carboxyhemoglobin (COHb) loading in the blood of rodents [31]. As a result, times to incapacitation of rodents by CO are longer in the presence of HCl. A further effect of the HCl is to cause a very rapid blood acidosis which, by shifting the oxygen-oxyhemoglobin equilibrium, results in oxygen being made more available to the animal [32]. Thus, the rodent may be enabled to resist incapacitation somewhat longer. (The decrease in RMV in rodents due to the presence of an irritant has not been observed with primates, where the effect is actually to increase the RMV. Furthermore, primates exhibit significantly decreased blood oxygen (PO_2) in the presence of the irritant, HCl [34].)

Overall, the effect of the irritant, HCl, with rats is to delay incapacitation by CO; however, it also increases the incidence of postexposure deaths due to the combined insult. This latter effect results in an apparent increase in the toxicity of the CO. These same effects are seen with exposure of rats to smoke from polyvinyl chloride when CO is also present [32].

Carbon dioxide is quite low in its own toxicological potency and is not, by itself, normally considered to be a significant factor in combustion toxicology. However, it does stimulate respiration, leading to an increased rate of loading of blood COHb from inhalation of CO. The same equilibrium COHb saturation is reached as in the absence of CO_2; however, times to incapacitation are generally longer. This may be due to the oxygen-oxyhemoglobin equilibrium shift, comparable to that caused by HCl. Increased incidence of lethality (particularly post-exposure) has been observed with certain combinations of CO and CO_2 [35], perhaps again analogous to that seen with mixtures of CO and HCl. The effect may be associated with the combined insult of respiratory acidosis (caused by CO_2) and metabolic acidosis (caused by CO), a condition from which the rodent has difficulty recovering post-exposure. Whether or not these effects of CO_2 occur with primates has not been determined.

Models incorporating combined effects of CO, CO_2, O_2, HCN and HCl are being worked out, but little validation data have been published. Methodology has been applied to a very limited extent to the prediction of LC_{50} values in the laboratory testing of materials for smoke toxicity [27,28]. Results, based only on CO and HCN analyses, have been quite promising. Current work on adding other toxicants to an FED or similar model should enable considerable "testing" to be conducted without the use of live animals. Ramifications resulting from models for toxicant mixtures are beginning to suggest that real smoke atmospheres may be much more toxic than one would initially suspect from consideration of the concentrations of individual toxicants taken separately.

The FED methodology has also been extended to real smoke or fire effluent atmospheres (even with actual toxicant concentrations being unknown), with exposure doses being expressed as mass of fuel consumed per unit volume multiplied by the exposure time (e.g., mg • L^{-1} • min) [7]. Summation or integration of incremental exposure doses can be made to apply to real smoke as well as to single toxicants. This is now being done as the modeling of toxicological hazards advances to reality [36].

REFERENCES

1. Ferguson, G. E. "Fire Gases," *Quarterly of the National Fire Protection Association*, 27(2):110 (1933).
2. Zapp, J. A. "The Toxicology of Fire," Medical Division Special Report No. 4, U.S. Army Chemical Center, MD (1951).
3. Kimmerle, G. "Aspects and Methodology for the Evaluation of Toxicological Parameters During Fire Exposure," *J. Fire and Flammability/Combustion Toxicology Supplement*, 1:42 (February 1974).
4. Birky, M. M. "Philosophy of Testing for Assessment of Toxicological Aspects of Fire Exposure," *J. Combustion Toxicology*, 3:5–23 (1976).

5. Birky, M. M. "Hazard Characteristics of Combustion Products in Fires: The State-of-the-Art Review," NBSIR 77-1234, U.S. National Bureau of Standards, Washington, D.C. (1977).

6. "Fire Toxicology: Methods for Evaluation of Toxicity of Pyrolysis and Combustion Products," Report No. 2, Committee on Fire Toxicology, The National Research Council, National Academy of Sciences, Washington, D.C. (1977).

7. Kaplan, H. L., A. F. Grand and G. E. Hartzell. *Combustion Toxicology: Principles and Test Methods*, Lancaster, PA:Technomic Publishing Co., Inc. (1983).

8. "Fire and Smoke: Understanding the Hazards," Committee on Fire Toxicology, The National Research Council, National Academy of Sciences, Washington, D.C. (1986).

9. "Fire Research on Cellular Plastics: The Final Report of the Products Research Committee," J. W. Lyons, Chairman, Products Research Committee, National Bureau of Standards, Washington, D.C., p. 5–9 (1980).

10. "Determination of the Toxicity Index of the Products of Combustion from Small Specimens of Materials," Naval Engineering Standard 713, British Ministry of Defense, London, England.

11. German Standards Institution. DIN 53436, "Erzeugung thermischer Zersetzungsprodukte von Werkstoffen unter Luftzufuhr und ihre toxikologische Prüfung. Part 1, Zersetzungsgerat und Bestimmung der Versuchstemperatur. Part 2, Verfahren zur thermischen Zersetzung" (January 1979).

12. Levin, B. C., A. J. Fowell, M. M. Birky, M. Paabo, A. Stolte and D. Malek. "Further Development of a Test Method for the Assessment of the Acute Inhalation Toxicity of Combustion Products," NBSIR 82-2532, National Bureau of Standards, Washington, D.C. (1982).

13. Alexeeff, G. V. and S. C. Packham. "Use of a Radiant Furnace Model to Evaluate Acute Toxicity of Smoke," *J. Fire Sciences*, 2(4):306–320 (July/ August 1984).

14. Kaplan, H. L., A. F. Grand and G. E. Hartzell. *Combustion Toxicology: Principles and Test Methods*, Lancaster, PA:Technomic Publishing Co., Inc., pp. 83–108 (1983).

15. "Combustion Toxicity Testing," New York State Uniform Fire Prevention and Building Code, Article 15, Part 1120, Office of Fire Prevention and Control, Department of State, New York State, Albany, NY (1986).

16. Tsuchiya, Y. "New Japanese Standard Test for Combustion Gas Toxicity," *J. Combustion Toxicology*, 4:5–7 (February 1977).

*17. Debanne, S. M., H. S. Haller and D. Y. Rowland. "A Statistical Comparison of Test Protocols Used in the Assessment of Combustion Product Toxicity," *J. Fire Sciences* (in press).

18. Hirschler, M. M. "Fire Hazard and Toxic Potency of the Smoke From Burning Materials," *J. Fire Sciences*, 5(5):289–307 (September/October 1987).

19. Babrauskas, V. (personal communication).

20. Tsuchiya, Y. and K. Sumi. "Evaluation of the Toxicity of Combustion Products," *J. Fire and Flammability*, 3:46–50 (1972).

21. Armstrong, G. W. "A Chemical/Mathematical Model for Predicting the Potential Physiological Hazard of a Changing Fire Environment," *J. Fire and Flammability/Combustion Toxicology*, 3:46–50 (1972).

22. Packham, S. C. and G. E. Hartzell. "Fundamentals of Combustion Toxicology in Fire Hazard Assessment," *J. Testing and Evaluation*, 9(6):341 (1981).

23. MacFarland, H. N. "Respiratory Toxicology," in *Essays in Toxicology*, 7, W. Hayes, Jr., ed. Academic Press, pp. 121-154 (1976).

*24. Alexeeff, G. V. and S. C. Packham. "Evaluation of Smoke Toxicity Using Concentration-Time Products," *J. Fire Sciences*, 2:362-379 (September/October 1984).

*25. Huggett, C. "The Forum: Reporting Combustion Product Toxicity Test Results," *J. Fire Sciences*, 3(2):79-82 (March/April 1985).

*26. Hartzell, G. E., S. C. Packham, A. F. Grand and W. G. Switzer. "Modeling of Toxicological Effects of Fire Gases: II. Mathematical Modeling of Intoxication of Rats by Carbon Monoxide and Hydrogen Cyanide," *J. Fire Sciences*, 3(2):115-128 (March/April 1985).

*27. Hartzell, G. E., W. G. Switzer and D. N. Priest. "Modeling of Toxicological Effects of Fire Gases: V. Mathematical Modeling of Intoxication of Rats by Combined Carbon Monoxide and Hydrogen Cyanide Atmospheres," *J. Fire Sciences*, 3(5):330-342 (September/October 1985).

28. Levin, B. C., M. Paabo, J. L. Gurman and S. E. Harris. "Effects of Exposure to Single or Multiple Combinations of the Predominant Toxic Gases and Low Oxygen Atmospheres Produced in Fires," *Fundamental and Applied Toxicology*, 9:236-250 (1987).

29. Kaplan, H. L. and G. E. Hartzell. "Modeling of Toxicological Effects of Fire Gases: I. Incapacitating Effects of Narcotic Gases," *J. Fire Sciences*, 2:286-305 (July/August 1984).

30. Hartzell, G. E., S. C. Packham, A. F. Grand and W. G. Switzer. "Modeling of Toxicological Effects of Fire Gases: III. Quantification of Post-Exposure Lethality of Rats from Exposure to HCl Atmospheres," *J. Fire Sciences*, 3(3):195-207 (May/June 1985).

*31. Hartzell, G. E., H. W. Stacy, W. G. Switzer, D. N. Priest and S. C. Packham. "Modeling of Toxicological Effects of Fire Gases: IV. Intoxication of Rats by Carbon Monoxide in the Presence of an Irritant," *J. Fire Sciences*, 3(4):263-279 (July/August 1985).

*32. Hartzell, G. E., A. F. Grand and W. G. Switzer. "Modeling of Toxicological Effects of Fire Gases: VI. Further Studies on the Toxicity of Smoke Containing Hydrogen Chloride," *J. Fire Sciences* (in press).

*33. Tsuchiya, Y. "On the Unproved Synergism of the Inhalation of Fire Gas," *J. Fire Sciences*, 4(5):346-354 (September/October 1986).

34. Kaplan, H. L., A. Anzueto, W. G. Switzer and R. K. Hinderer. "Effects of Hydrogen Chloride on Respiratory Response and Pulmonary Function of the Baboon," *J. Tox. Env. Health* (in press).

35. Levin, B. C., M. Paabo, J. L. Gurman, S. E. Harris and E. Braun. "Toxicological Interactions Between Carbon Monoxide and Carbon Dioxide," *Proceedings of 16th Conference of Toxicology*, Air Force Aerospace Medical Research Laboratory, Dayton, OH (October 1986).

36. Bukowski, R. W., W. W. Jones, B. M. Levin, C. L. Forney, S. W. Stiefel, V. Babrauskas, E. Braun and A. J. Fowell. *Hazard 1. Volume 1: Fire Hazard Assessment Method*, NBSIR 87-3602, U.S. Department of Commerce, National Bureau of Standards (July 1987).

NOTE: References noted with asterisk are contained in *Advances in Combustion Toxicology*.

3

Understanding of Hazards to Humans

G. E. HARTZELL

Department of Fire Technology
Southwest Research Institute
P.O. Drawer 28510
San Antonio, Texas 78284

MICHAEL FARADAY, DURING his lectures in 1848, was referring to the chemistry of a candle when he said,

> . . . There is not a law under which any part of this universe is governed which does not come into play, and is not touched upon, in these phenomena [1].

Fire is an exceedingly complex phenomenon encompassing physics, chemistry, heat transfer, fluid mechanics and gas dynamics. The field of combustion toxicology adds the complexities of many of the physiological and biochemical mechanisms of the respiratory, cardiovascular and central nervous systems of the human body. Thus, one should appreciate the dilemma of the combustion toxicologist, who must combine all these technologies, along with the field of materials science, into a meaningful assessment of the hazard and risk of human exposure to fire. This is no small task, particularly when complicated further by political, marketing, and even legal interests which accompany technology that deals with hazards to the human population.

In order to achieve a meaningful understanding of the toxic hazards to humans exposed to smoke from a fire, it is essential to determine what constitutes an exposure dose hazardous to human subjects. To obtain the necessary toxicological data, the combustion toxicologist must turn to laboratory experiments using animals exposed under conditions simulating those of fire atmospheres. This must be done since most available toxicological data relevant to humans have been developed specifically for assessment of long-term exposures in the work

19

place environment. The data are generally not applicable for evaluation of the acute effects from brief exposures to the relatively high concentrations of combustion products which may be present in a fire. Development of the desired toxicological data from scientific experiments under controlled laboratory conditions usually cannot be accomplished with human subjects since the experiments themselves would be hazardous. Research in this field must normally be limited to the exposure of experimental animals, usually rodents, or in some cases, nonhuman primates. Reasonably reliable extrapolation to human exposures may then be made, providing that both qualitative and quantitative differences in toxicological effects between laboratory animals and man are understood. Considerable progress has been made in such extrapolations through research studies involving exposure of laboratory animals, including nonhuman primates, to individual fire gas toxicants [2]. Results of these studies, particularly those involving carbon monoxide, hydrogen cyanide and hydrogen chloride, have led to a better understanding of the complex toxicities of smoke produced from the burning of materials and have opened the way toward the modeling of toxic hazards to humans exposed in fires.

Before a discussion of the effects of individual and multiple toxicants, it should be emphasized that there has been some tendency in the fire hazard community to oversimplify the intoxication of humans by inhalation of fire effluents. Only recently have serious efforts been made to sort out and understand the components of this very complex process such that hazards to humans can be estimated [2,3].

First, the accumulation of a body burden of a toxicant through inhalation is greatly affected by the physical activity of the subject. The rate and depth of respiration, expressed as the respiratory minute volume (RMV), may change as much as twenty-fold over the range of physical activity from "at rest" to "strenuous activity." Increased physical activity is also attended by an increased need for oxygen (oxygen demand) by critical body organs, especially the heart. Recent studies using guinea pigs have shown incapacitation from carbon monoxide to occur at one-half the concentration and twice as fast in running animals as compared to sedentary animals [4]. Thus, a person at rest or sleeping may be able to tolerate a fairly high blood carboxyhemoglobin (COHb) content simply because the body at rest has a relatively low oxygen demand. The same person, upon awakening and attempting to escape from a fire, may quickly lose consciousness due to the suddenly increased requirement for oxygen that his blood cannot supply. Further, the effect of panic on both RMV and oxygen demand is not well understood and is likely to be highly variable among individuals.

A second difficulty involves the fact that the major fire effluent toxicants are not only taken into the body but are simultaneously released and/or detoxified, thus further complicating assessment of the actual body burden. This likely accounts for the observation that relatively high exposure doses can be better tolerated if the concentration of a toxicant is low over a long exposure time.

A third, and often overlooked, element in combustion toxicology is that laboratory data are obtained using experimental animals of quite uniform age, body weight, and overall state of health. Humans, who might be exposed in a fire, may vary over the entire spectrum of the population, with most probably suffering from some degree of pre-existing compromised state of health. (The high incidence of significant blood alcohol content of fire victims is well documented [5,6].)

Overall, the extrapolation of laboratory animal data to predict hazard to humans must be done with extreme caution, particularly as to what might constitute the difference between a "safe" exposure and a "hazardous" one. An exposure which is safe to one person may kill another. Even in view of all these complexities, combustion toxicologists have reached some degree of general agreement as to what constitutes a hazardous exposure dose of some of the major fire effluent toxicants.

CARBON MONOXIDE

Carbon monoxide (CO), although not the most toxic of gases, is always one of the most abundant and, therefore, is the major threat in most fire atmospheres. Carbon monoxide exerts its toxicity by combining with the hemoglobin of the red blood cells to form carboxyhemoglobin (COHb). The affinity of hemoglobin for CO is more than 200 times greater than for oxygen. Even partial conversion of hemoglobin to COHb reduces the oxygen-transport capability of the blood, thereby resulting in a decreased supply of oxygen to critical body organs, such as the brain and heart. In addition, CO causes oxyhemoglobin to bond more tightly to its oxygen content than otherwise would occur, further decreasing the availability of oxygen to body tissues. The toxic effects that result from exposure to CO are primarily due to anemic hypoxia and depend on the concentration of the CO and the duration of the exposure. Effects of CO resulting from exposure to various concentrations for different times are summarized in Table 1. Other factors, such as the rate of uptake of the CO and the state of health of the individual, also influence the resulting effects.

Although many factors may cause greater susceptibility, the symptoms produced by exposure to CO are directly related to the amount of hemoglobin of the blood that is converted to COHb. Carboxyhemoglo-

Table 1. Effects of CO in man [7].

Concentration (ppm)	Effects
200	Possibly headache, mild frontal in 2 to 3 hours
400	Headache, frontal, and nausea after 1 to 2 hours; occipital after 2-1/2 to 3-1/2 hours
800	Headache, dizziness and nausea in 3/4 hours; collapse and possibly unconsciousness in 2 hours
1,600	Headache, dizziness and nausea in 20 minutes; collapse, unconsciousness, possibly death in 2 hours
3,200	Headache and dizziness in 5 to 10 minutes; unconsciousness and danger of death in 30 minutes
6,400	Headache and dizziness in 1 to 2 minutes; unconsciousness and danger of death in 10 to 15 minutes
12,800	Immediate effect, unconsciousness and danger of death in 1 to 3 minutes

bin can readily be measured in a clinical laboratory and is expressed as percent COHb saturation. Symptoms observed in humans at various COHb blood saturation percentages are shown in Table 2. These data indicate that, at COHb levels below 30 percent, effects are normally not sufficiently severe as to impair human escape capability and, at levels above 40 percent, escape may be severely impaired or even not possible. Thus, the COHb blood saturation level that would impair escape of many individuals appears to be in the range of 30 to 40 percent. At COHb levels above 60 percent, very severe symptoms occur and death is likely. Although some believe that death can be attributed solely to CO only when the COHb level is 60 percent or greater [9], a figure of 50-percent COHb would seem to be more appropriate. Support for this criterion comes from analysis of data from victims of improperly operated gas heaters which showed a mean COHb saturation of 49.5 percent with a standard deviation of 14.0 percent [10].

In an individual exposed to CO, the effects of COHb blood saturation depend not only on the CO concentration and duration of exposure but also on the pulmonary ventilation of the individual. The greater the pulmonary ventilation, the more rapid will be the uptake of CO by the blood until an equilibrium level is reached. The extent of pulmonary ventilation is determined by the respiratory rate and the tidal volume.

At rest, the respiratory minute volume (RMV) of man averages approximately 5 L/min, but may increase during strenuous activity to as much as 100 L/min. Concentration-time curves showing the effects of CO concentration and duration of exposure on COHb loading by man with an assumed RMV of 20 L/min (light activity) were derived using the Stewart-Peterson equation [11] and are shown in Figure 1 [2]. From the curves corresponding to 30 to 40-percent COHb, escape impairment of humans engaged in light physical activity would be anticipated after inhalation of an accumulated concentration-time (Ct) exposure dose of CO in the range of approximately 35,000 to 45,000 ppm-min. Under conditions of strenuous physical activity, the RMV could increase severalfold, resulting in a more rapid loading of COHb and in incapacitation or death after a much smaller accumulated Ct exposure dose.

Nonhuman primate studies have shown incapacitation (escape impairment) to occur at CO exposure doses of about 34,000 ppm-min (5-minute exposures) [2,12]. Thus, the evidence appears to suggest a simple rule-of-thumb for estimation of CO exposure doses likely to be hazardous to humans. Any exposure in which the product of concentration (ppm) × time (minutes) exceeds about 35,000 ppm-min is likely to be dangerous [2]. For example, a 10-minute exposure to 3500 ppm of

Table 2. Human response to various levels of blood carboxyhemoglobin saturation [8].

COHb Saturation (percent)	Symptoms in Humans
0–10	None
10–20	Tension in forehead, dilation of skin vessels
20–30	Headache, pulsation in sides of head
30–40	Severe headache, ennui, dizziness, weakening of eyesight, nausea, vomiting, prostration
40–50	Same as above, increase in breathing rate and pulse, asphyxiation and prostration
50–60	Same as above, coma, convulsions, Cheyne-Stokes respiration
60–70	Coma, convulsions, weak respiration and pulse, death possible
70–80	Slowing and stopping of respiration, death within hours
80–90	Death in less than an hour
90–100	Death within a few minutes

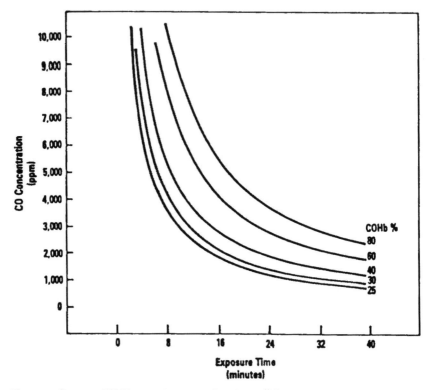

Figure 1. Percent COHb as a function of ambient CO concentration and exposure time for humans with an RMV of 20 L/min. Based on Stewart-Peterson Equation: Log (Δ% COHb/L) = 1.036 Log (CO concentration)—4.4793 [11].

carbon monoxide would be expected to be hazardous and possibly incapacitating to many people. The concentration × time rule-of-thumb must be applied with caution at high concentrations, since progressively lower exposure doses can be tolerated as the concentration is increased. It is, however, reasonably applicable for the range of carbon monoxide concentrations normally generated in fires.

Although overt symptoms do not generally occur at COHb saturations of less than 10 percent, low levels may not be without adverse effects. For example, exposure to CO sufficient to produce COHb saturations in the 3 to 5-percent range has been reported to impair cardiovascular function in patients with cardiovascular disease [13,14] as well as in normal subjects [15,16]. Carboxyhemoglobin saturations in the 4 to 6-percent range have been shown to significantly reduce the threshold for ventricular fibrillation both in normal, anesthetized dogs and also in those with acute myocardial injury [17]. These levels also increase ischemia associated with acute myocardial infarction in dogs [18].

The brain also requires oxygen for normal functions and is sensitive to a diminished supply. Several human experimental studies have been conducted in an effort to determine the behavioral effects of low concentrations of CO [19]. Evidence has been obtained that COHb saturations in the 3 to 5-percent range may adversely affect man's ability to detect small unpredictable environmental changes, i.e., impair vigilance capability. It also appears that an abrupt elevation of COHb saturation to 5 percent will transiently alter the visual light threshold. In other studies, low concentrations of CO have been reported to produce adverse effects on reaction time and time discrimination, but whether significant changes in mental functions occur at COHb concentrations between 2 and 5 percent is controversial. At concentrations of COHb below 10 percent, the ability to perform complex tasks requiring both judgement and motor coordination is not adversely affected.

HYDROGEN CYANIDE

Hydrogen cyanide (HCN) is one of the most rapidly acting toxicants [20], as well as being approximately 20 times more toxic than carbon monoxide. Inhalation of HCN may cause severe toxic effects and death from within a few minutes up to several hours, depending on the concentration inhaled. The action of HCN is due to the cyanide ion, which is formed by hydrolysis in the blood. Unlike CO, which remains primarily in the blood, cyanide ion is distributed throughout the body water and is in contact with cells of tissues and organs. Cyanide reacts readily with trivalent iron of the cytochrome oxidase enzyme to form a cytochrome oxidase-CN complex and also with trivalent iron of methemoglobin to form cyanomethemoglobin. Since cytochrome oxidase occupies a central role in the utilization of oxygen in practically all cells, its inhibition rapidly leads to loss of cellular functions (cytotoxic hypoxia), then to cell death. In contrast with CO, cyanide does not decrease the availability of oxygen but, rather, prevents the utilization of oxygen by the cells. The heart and brain are particularly susceptible to this inhibition of cellular respiration, with bradycardia, cardiac arrhythmias and EEG brain wave activity indicative of central nervous system depression having been reported in studies using monkeys [21]. Although such cardiac irregularities are often noted in HCN intoxication, the heart invariably outlasts respiration and death is usually due to respiratory arrest of central nervous system origin.

If the concentration of cyanide ion is not sufficiently great as to cause death, it is slowly released from its complexes with cytochrome oxidase and methemoglobin and converted to thiocyanate ion by the enzyme, rhodanese. Treatment for cyanide poisoning is based on the competition between cytochrome oxidase and methemoglobin for cyanide. Nitrites are administered in order to produce a high concentration of methemoglobin in the blood. Methemoglobin competes with cytochrome oxidase for cyanide, reacting to form cyanomethemoglobin. Actual detoxification is then achieved by the administration of thiosulfate, which, under the influence of rhodanese, reacts with cyanide to form thiocyanate. Thiocyanate is a relatively nontoxic species and is readily excreted in the urine.

Data relating symptoms in humans and also in nonhuman primates to various concentrations of HCN are very limited. One widely used source of descriptive accounts of hydrogen cyanide intoxication of humans reports that 50 ppm may be tolerated by man for 30 to 60 minutes without difficulty; 100 ppm for that same period is likely to be fatal; 135 ppm may be fatal after 30 minutes; and 181 ppm may be fatal after 10 minutes [8]. Since incapacitation normally occurs at one-third to one-half the lethal exposure dose, the data suggest that exposure doses for incapacitation by HCN range from approximately 2500 ppm-min at 100 ppm to 750 ppm-min at 200 ppm. Using a rule-of-thumb analogous to that for carbon monoxide, it would appear that a product of hydrogen cyanide concentration (ppm) × time (minutes) in the range of about 1500 ppm-min would likely be hazardous to humans [2]. Experiments with monkeys would appear to confirm this value, with exposure doses in the range of 1200 to 1900 ppm-min being reported as incapacitating [21]. With hydrogen cyanide, in particular, progressively lower exposure doses can be tolerated as the concentration is increased. Therefore, the concentration × time rule-of-thumb must be applied with great caution at high concentrations. It is, however, reasonably applicable for concentrations of hydrogen cyanide generally found present in fire atmospheres.

As a complicating effect, cyanide ion has been reported to stimulate respiration by its action on the chemoreceptors of the carotic and aortic bodies [20]. It is hypothesized that cyanide blocks cellular respiration in the chemoreceptor cells, which then respond as they do to decreased oxygen in the blood by stimulation of the medullary respiratory center. This hyperventilatory effect has been proposed as a possible mechanism whereby HCN causes a more rapid uptake of CO in fire victims exposed to smoke containing both CO and HCN.

In contrast to the determination of blood COHb levels, measurement of cyanide in the blood (or tissues) has been less definitive in establishing HCN intoxication. There are a number of possible reasons for the somewhat limited usefulness of blood cyanide data. One reason is that there does not appear to be a correlation between exposure concentrations of HCN and blood cyanide concentrations. In two animal studies, investigators reported that blood cyanide levels of rats exposed to HCN did not correlate with atmospheric HCN concentrations and that it was not possible to define a consistent blood loading curve for cyanide [22,23]. In other studies, in which animals were exposed to HCN until death occurred, there was considerable variability in cyanide levels in blood and tissues at death [24]. In a preliminary study of HCN toxicity conducted at Southwest Research Institute, the data, although limited, showed greater consistency in blood cyanide loading curves; however, blood cyanide levels reached approximately the same maximum concentration regardless of the HCN exposure concentration [25]. On the other hand, studies with monkeys exposed to HCN showed that blood cyanide levels were linearly correlated with HCN exposure concentrations, providing corrections were made for differences in RMV for different animals [21].

The inconsistencies and variability in blood cyanide levels that have been observed in animal studies may be at least partially due to analytical methodologies used in these studies and/or to changes in cyanide concentrations under certain storage conditions. It is also possible that the difficulties in analysis of cyanide in blood and the storage effects are responsible for differences among investigators in what are considered "normal" and toxic blood cyanide levels. An early study reported background levels of 0 to 0.14 μg/mL of cyanide in the blood [26]. More recently, it was reported that normal blood cyanide values for humans range from 0 to 0.22 μg/mL, with an average of 0.05 μg/mL for non-smokers, and that the higher levels of this range were due to cigarette smoking [27]. In the report of a fire fatality study [28], the investigators considered 0 to 0.25 μg/mL cyanide in the blood as "normal," 0.26 to 1.00 μg/mL as "subtoxic," 1.00 to 2.00 μg/mL as "possibly toxic," and >2.00 μg/mL as "probably toxic." A further study [29] considered 3.00 μg/mL as the minimum lethal level of blood cyanide, while still another [30] reported that inhalation of HCN produces symptoms at concentrations at or above 0.2 μg/mL of cyanide in the blood and that, in acute cyanide poisoning, the blood level is likely to be 5.0 μg/mL. In any event, blood cyanide analyses should be highly suspect in establishing HCN as contributing to a smoke inhalation death. The analytical method used, sample storage history and reputation of the laboratory should all be carefully examined.

INSUFFICIENT OXYGEN

Oxygen (O_2) is consumed from the atmosphere during combustion. When oxygen drops from its usual level of 21 percent in air to approximately 17 percent, a person's motor coordination is impaired. When oxygen reaches the range of 14 to 10 percent, a person is still conscious but may exercise faulty judgement and will be quickly fatigued. In the range of 10 to 6 percent oxygen, a person loses consciousness and must be revived with fresh air or oxygen within a few minutes to prevent death [31]. During periods of exertion, increased oxygen demand may result in oxygen deficiency symptoms at higher oxygen levels.

It is the opinion of this author that low oxygen hypoxia is probably not a particularly significant factor in estimating toxic hazards to humans exposed to fire. Oxygen levels normally remain quite high in the fuel-controlled developing fire, with a very rapid drop only as the fire passes into a ventilation-limited state at and beyond flashover. At this point, the circumstances are untenable to humans for numerous other reasons. Oxygen levels are usually near normal at locations sufficiently remote from the fire that heat is not a problem. However, it is possible to propose specific scenarios in which oxygen depletion may be a significant factor.

CARBON DIOXIDE

Carbon dioxide (CO_2) is usually evolved in large quantities from fires. While not particularly toxic at observed levels, moderate concentrations of carbon dioxide increase both the rate and depth of breathing, thereby increasing the RMV. This condition contributes to the overall hazard of a fire gas environment by causing accelerated inhalation of toxicants and irritants. The rate and depth of breathing are increased about 50 percent by 2-percent carbon dioxide. If 4-percent carbon dioxide is breathed, the RMV is approximately doubled, but the effect may be scarcely noticed by the individual. Any further increase in carbon dioxide from 4 percent up to 10 percent produces a correspondingly greater RMV, and at 10 percent, the RMV may be 8 to 10 times the resting level. The subject also may have symptoms of dizziness, faintness, and headache [32].

HYDROGEN CHLORIDE

Hydrogen chloride (HCl) is formed from the combustion of chlorine containing materials, the most notable of which is polyvinyl chloride

(PVC). It is both a potent sensory irritant and also a strong pulmonary irritant. It is a strong acid, being corrosive to sensitive tissue such as the eyes. Concentrations as low as 75 ppm are extremely irritating to the eyes and the upper respiratory tract, and behavioral impairment has been suggested [33]. However, hydrogen chloride has been found not to be physically incapacitating to nonhuman primates subjected to concentrations up to 17,000 ppm for 5 minutes [12]. The toxicant was reported, however, to cause an incidence of postexposure death at these high doses. Comparable studies have not been conducted using actual PVC smoke, with which it is claimed that other irritants also may be present [33]. There are also questions as to the extent of any respiratory dysfunction and susceptibility to infection caused by exposure to hydrogen chloride and PVC smoke. One study using baboons exposed to PVC smoke containing up to 4000-ppm HCl did not indicate any significant residual effects on pulmonary function when tested at 3-days and 3-months postexposure [34].

Extrapolation of rodent data to humans is complicated by species differences even among rodents in their sensitivity to HCl, as well as to other irritant gases. A comparison of RD_{50} (respiratory depression) and LC_{50} values of HCl demonstrates that the mouse is more sensitive than the rat to both the sensory irritant and also to the lethal effects of HCl. As sensitive as the mouse is, it has been claimed that the mouse may still be 7 to 10 times less sensitive than man and that a correction factor is required to extrapolate mice lethality data to man [35]. The correction factor is based on the observation that HCl or smoke from PVC is about 7 to 10 times more lethal in mice exposed via a tracheal cannula rather than through normal nose inhalation [36]. The assumption was made that the respiratory tract of the mouse fitted with a tracheal cannula is comparable to that of a mouth-breathing human, an assumption which may be of questionable validity. It ignores the scrubbing capability of man's oral mucosa and marked differences in the tracheobronchial and pulmonary regions between the mouse and man [37]. Furthermore, studies with the baboon, an animal species whose upper airway is recognized as a useful model for the human [38], refute the claim that man is more sensitive than the mouse to the lethal effects of HCl. These studies have shown unquestionably that primates can survive short exposures to much higher concentrations of HCl than can the mouse [12]. However, man may well be more sensitive than rodents in terms of sensory irritation, an effect that is probably more significant in terms of escape from a fire than is postexposure lethality.

Studies of the respiratory effects of both HCl and PVC smoke have been conducted, with marked differences being observed between rodents and nonhuman primates [34,39]. In rodents (mice, rats and guinea pigs), the response to both HCl and PVC smoke consists of a decrease in respiratory rate and minute volume, the typical response to a sensory irritant. In the anesthetized baboon, the responses both to PVC smoke and to HCl were similar, being characterized by an increase in respiratory rate and minute volume. Although blood oxygen (PaO_2) values were reduced during exposure to both flaming and nonflaming PVC smoke, the hypoxemia was not as severe as that produced by exposure to a comparable concentration of pure HCl gas. Also, changes in pH and $PaCO_2$ values in PVC smoke-exposed animals were less marked than in HCl-exposed animals.

Considerable controversy exists as to what concentration of HCl is hazardous to man. Although numerous studies of the acute effects of HCl have been conducted with rodents, it is questionable whether lethality data from rodents can be directly extrapolated to man because of anatomical differences in the respiratory tract of the rodent and the primate. However, exposure doses (concentration × time) of HCl which cause postexposure lethality in rats are in the same range as those which have resulted in postexposure deaths of baboons, although the data for baboons are very limited and the comparison made is rather subjective [40]. The lethal toxic potency of HCl with rats is actually only somewhat greater than that of carbon monoxide [41]. Consideration of the exposure dose of carbon monoxide thought to be *hazardous* to humans would lead one, based simply on relative lethal toxic potencies, to suspect that exposure of humans to the range of 700 ppm or more of HCl for 30 minutes would be highly dangerous.

ATMOSPHERES CONTAINING MULTIPLE TOXICANTS

The consequences of exposure to atmospheres containing multiple toxicants have only recently begun to be realized. Although individual fire gas toxicants may exert quite different physiological effects through different mechanisms, when present in a mixture each may result in a certain degree of compromise experienced by an exposed subject. It should not be unexpected that varying degrees of a partially compromised condition may be roughly additive in contributing to incapacitation or death. This has been demonstrated in a number of studies, and is a key element in the assessment of toxic hazard from analytical data.

There is a further complication with combined toxicants which is more difficult to deal with. Each individual toxicant may also have physiological effects other than that of its principal specific toxicity.

These effects, particularly those involving the respiratory system, may alter the rate of uptake of other toxicants. For example, hydrogen cyanide is known to cause hyperventilation early in an exposure, with four-fold increases in the RMV of monkeys being reported [21]. (Respiration eventually slows as narcosis results, however.) It has been suggested that this initial hyperventilation in primates may result in faster incapacitation from HCN than would be expected, along with more rapid COHb saturation should CO also be present. Carbon monoxide, in decreasing the oxygen transport capability of the blood, eventually results in a condition of metabolic acidosis. A slowing of respiration also results from the induced state of narcosis. Carbon dioxide is a respiratory stimulant which increases the rate of uptake of other toxicants, thus producing a faster rate of formation of COHb from inhalation of CO. Irritants, such as HCl, slow respiration in rodents due to sensory irritation, but increase RMV in primates from pulmonary irritation. With primates, there is also evidence of an irritant (HCl) causing bronchoconstriction which may interfere with oxygen reaching the alveoli for diffusion into the blood. With all these effects possible in the inhalation of mixtures of toxicants in real fire effluents, the situation is extremely complex. Very little research using toxicant combinations has been conducted using primates and the full seriousness of the combined effects on incapacitation and death of humans exposed to fire gas atmospheres is not fully understood. It is reasonable to say, however, that the effects described certainly do not favor the safety of those exposed.

In spite of the complexity of dealing with atmospheres containing multiple toxicants, considerable progress has been made in understanding some of the effects from studies using rodents. For example, it is fairly well agreed that carbon monoxide and hydrogen cyanide appear to be additive when expressed as fractional doses required to cause an effect [42]. Thus, as a reasonable approximation, the fraction of an effective dose of CO can be added to that of HCN and the time at which the sum becomes unity (100 percent) can be used to estimate the presence of a hazardous condition.

In the case of mixtures of hydrogen chloride and carbon monoxide, empirical analysis of toxicological data shows that exposure doses leading to lethality of rats may also be additive [41]. Although not confirmed with primates, these studies imply that hydrogen chloride may be much more dangerous than previously thought when in the presence of carbon monoxide or, conversely, carbon monoxide intoxication may be much more serious in the presence of an irritant. Quite striking is the rapid respiratory acidosis seen in the blood of rats exposed to HCl, which when coupled with the metabolic acidosis produced by CO, results in severely compromised animals. This could be quite serious with real fires involving PVC, since carbon monoxide is almost always present from the burning of other materials. Observations with rodents also suggest complications immediately following exposure to atmospheres containing both hydrogen chloride and carbon monoxide [41]. This may have significance with human exposure, such as prolonged hypoxemic conditions following rescue or escape. There is also some suggestion that the incapacitating effects of carbon monoxide may be enhanced in primates upon simultaneous exposure to HCl, the presence of which causes the blood PaO_2 to be decreased [43]. This is probably the case with other irritants as well.

Studies using rats exposed to mixtures of CO_2 and CO have shown an incidence of postexposure deaths which would not be attributed to the CO [44]. The effect, although probably not one of toxicological additivity, may be associated with the combined effects of respiratory acidosis and metabolic acidosis, a condition from which the rodent has difficulty recovering. Any effect of CO_2 on increasing the hazard to humans due to CO exposure in fires is unknown. The more rapid uptake of CO from the increased RMV would seem to increase the hazard. However, it has been observed with rats that, although carbon dioxide increases the rate of uptake of carbon monoxide, it does not result in an equilibrium blood carboxyhemoglobin saturation significantly different from that obtained with the same concentration of carbon monoxide alone. This effect has not been studied using primates.

It was stated earlier that low oxygen hypoxia may not, in itself, be an important factor in estimating toxic hazards in fires. However, oxygen consumption in a fire atmosphere would normally be expected to be accompanied by the presence of a significant concentration of carbon monoxide. The effects of toxicants, such as carbon monoxide, in the presence of conditions of low oxygen hypoxia have not been studied to any great extent. Some work suggests significant interaction leading to a greater susceptibility of exposed subjects [44,45]. Studies using mice have also shown hydrogen cyanide to be unusually dangerous when coupled with low oxygen hypoxia [45].

It should be emphasized that in most real, full-scale, multi-material fires, exposure doses of CO produced normally far exceed those of other toxicants and that, to a good first approximation, rough tenability limits can be predicted solely on the basis of CO produced. However, significant combined effects of toxicants have certainly been demonstrated with rodents in the laboratory. These need to be studied further with nonhuman primates in order to determine their impact on hazards to humans. Smoke atmospheres may be much more hazardous than one would initially suspect from consideration of the concentrations of the individual toxicants taken separately.

ESTIMATION OF TOXIC HAZARD TENABILITY LIMITS FROM FIRE DATA [42]

Fully furnished full-scale room/corridor/remote room fire tests have been reported from Southwest Research Institute [46]. Tests were initiated with a smoldering fire in a chair located close to a sofa. Following flaming ignition of the chair (after about 19 minutes of smoldering) the fire progressed rapidly to flashover within an additional 8 minutes, accompanied by life-threatening conditions of visual obscuration, temperature, carbon monoxide, hydrogen cyanide and oxygen depletion in the room of origin. After flashover, tenability in a remote room also quickly deteriorated, first with visual obscuration by smoke, followed by rapidly increasing concentrations of toxic gases. Test animals (rats) placed in a remote room became incapacitated at 28.4 ± 1.4 minutes from initiation of the test, with death occurring at 36.6 ± 3.7 minutes. Using toxicant concentration data, with a Fractional Effective Dose model based only on carbon monoxide and hydrogen cyanide, predicted times to incapacitation and death of the rats were calculated to be 30.2 and 37.0 minutes, respectively [42]. (It has been suggested that accumulated exposure doses of carbon monoxide and hydrogen cyanide sufficient to incapacitate rats would likely be severely compromising to humans [2].)

SMOKE

Under circumstances for which fire gas toxicant concentration data are not available, tenability can still be estimated based on smoke exposure dose. Laboratory toxic potency data, obtained using rodents, are available for many materials [47]. A survey of these data reveals that smokes from most common materials are characterized by lethal exposure doses in the range of 300 mg-min/L to 1500 mg-min/L, with the average being about 900 mg-min/L. Hazardous, or possibly incapacitating, doses would be expected to lie at perhaps 300 mg-min/L. This latter value would correspond to about 9 kg-min/30 m^3, i.e., in a 3 × 4 × 2.5-m room.

Calculations based on this exposure dose show that exposure to smoke for 30 minutes from the burning of only about 300 g of fuel load in the 3 × 4 × 2.5-m room would be likely hazardous to an occupant.

It would also be predicted that the burning of only about 12 kg of a material, testing as "typical" in a smoke toxicity test, would produce sufficient smoke as to be hazardous in 20 such rooms upon only 15 minutes of exposure. It would appear quite understandable that a fire in one hotel room could easily prove fatal to occupants in other rooms. Smoke, even of "average" toxicity, is still very toxic!

THE IMPORTANCE OF TIME

Escape is the key to survival from a fire, with tenability being the time available for escape. The time factor has two components, only one of which is the time for an exposed person to accumulate a hazardous dose. Of greater importance is the time required to generate and transport the quantity of smoke which would result in a hazardous dose. Thus, tenability is very much a function of fire growth rate.

From the viewpoint of life safety, it is most important that a hazardous exposure dose of smoke either never be generated (rapid extinguishment) or be prevented from reaching building occupants (smoke management). These two factors of prevention are undoubtedly more significant in terms of life safety than the choosing of one material over another on the basis of smoke toxicity test data.

REFERENCES

1. Lyons, J. W. *Fire*, New York, NY:Scientific American Books, Inc., p. 7 (1985).
*2. Kaplan, H. L. and G. E. Hartzell. "Modeling of Toxicological Effects of Fire Gases: I. Incapacitating Effects of Narcotic Fire Gases," *J. Fire Sciences*, 2:286–305 (July/August 1984).
3. Purser, D. A. "Modeling Toxic and Physical Hazard in Fire," to be published in *J. Fire Sciences*.
4. Alarie, Y., M. Schaper, D. Malek and F. Esposito. "Toxicity of Plastic Combustion Products," Summaries of Center for Fire Research In-House Programs and Grants-1987, NBSIR 87-3650, U.S. Department of Commerce, National Bureau of Standards, pp. 45–46 (October 1987).
5. Harland, W. A. and W. D. Woolley. "Fire Fatality Study–University of Glasgow," Information Paper IP 18/79, Building Research Establishment, Glasgow, Scotland (1979).
6. Halpin, B. M. and G. W. Berl. "Human Fatalities from Unwanted Fires," Report NBS-GCR-79-168, National Bureau of Standards, Washington, D.C. (1978).
7. Hamilton, A. and H. L. Hardy. *Industrial Toxicology*. Third Edition, Acton, MA:Publishing Sciences Group (1974).
8. Kimmerle, G. "Aspects and Methodology for the Evaluation of Toxicological Parameters During Fire Exposure," *The Journal of Fire and Flammability Combustion Toxicology Supplement*, 1:42 (1974).

9. Birky, M. M., M. Paabo and J. E. Brown. "Correlation of Autopsy Data and Materials Involved in the Tennessee Jail Fire," *Fire Safety Journal*, 2:17–22 (1979/80).

10. Packham, S. C. "Forensic Applications of Combustion Toxicology," *Proceedings, Toxic Hazards from Fire Workshop*, Washington, D.C., November 15–16, 1984, Lancaster, PA:Technomic Publishing Company (1984).

11. Stewart, R. D., J. E. Peterson, T. N. Fisher, M. J. Hosko, E. D. Baretta, H. C. Dodd and A. A. Herrmann. "Experimental Human Exposure to High Concentrations of CO," *Arch. Env. Health*, 26:1–7 (1973).

*12. Kaplan, H. L., A. F. Grand, W. G. Switzer, D. S. Mitchell, W. R. Rogers and G. E. Hartzell. "Effects of Combustion Gases on Escape Performance of the Baboon and the Rat," *J. Fire Sciences*, 3(4):228–244 (July/August 1985).

13. Ayers, S. M., S. Giannelli, Jr. and H. Mueller. "Effects of Low Concentrations of Carbon Monoxide. Part IV. Myocardial and Systemic Responses to Carboxyhemoglobin," *Ann. N. Y. Acad. Sci.*, 174:268–293 (1970).

14. Aronow, W. S., E. A. Stemmer and M. W. Isbell. "Effect of Carbon Monoxide Exposure on Intermittent Claudication," *Circulation*, 49:415–417 (1974).

15. Drinkwater, B. L., P. B. Raven, S. M. Harvath, J. A. Gliner, R. O. Ruhling, N. W. Boldaun and S. Taguchi. "Air Pollution, Exercise, and Heat Stress," *Arch. Env. Health*, 28:177–181 (1974).

16. Aronow, W. S. and J. Cassidy. "Effect of Carbon Monoxide on Maximal Treadmill Exercise," *Ann. Intern. Med.*, 83:496–499 (1975).

17. Aronow, W. S., E. A. Stemmer and S. Zweig. "Carbon Monoxide and Ventricular Fibrillation Threshold in Normal Dogs," *Arch. Env. Health*, 34:184–186 (1979).

18. Becker, L. C. "Augmentation of Myocardial Ischemia by Low Level Carbon Monoxide Exposure in Dogs," *Arch. Env. Health*, 34:274–279 (1979).

19. Medical and Biological Effects of Environmental Pollutants: Carbon Monoxide, National Academy Sciences, Washington, D.C. (1977).

20. Swinyard, E. A. "Noxious Gases and Vapors: Carbon Monoxide, Hydrocyanic Acid, Benzene, Gasoline, Kerosene, Carbon Tetrachloride, and Miscellaneous Organic Solvents," *The Pharmacological Basis of Therapeutics*, Third Edition, Edited by Louis Goodman and Alfred Gilman, Chapt. 44, pp. 915–928 (1965).

21. Purser, D. A., P. Grimshaw and K. R. Berrill. "Intoxication by Cyanide in Fires: A Study in Monkeys Using Polyacrylonitrile," *Arch. Env. Health*, 39:394 (1984).

22. Birky, M. M., M. Paabo, B. C. Levin, S. E. Womble and D. Malek. "Development of Recommended Test Method for Toxicological Assessment of Inhaled Combustion Products," NBSIR 80-2077, National Bureau of Standards, Washington, D.C. (September 1980).

23. Farrar, D. G., W. A. Galster and B. M. Hughes. "Toxicological Evaluation of Material Combustion Products," UTECH 79-141, The University of Utah, Salt Lake City (November 1979). (An appendix to this report is entitled, "Sensitivity of the Leg-Flexion Avoidance and Loading of Arterial Blood in Rats Exposed to Hydrogen Cyanide Gas," by W. A. Galster.)

24. Smith, P. W. "FAA Studies of the Toxicity of Products of Combustion," *Proceedings of the First Conference and Workshop on Fire Casualties*, B. M. Halpin, ed. The Johns Hopkins University, Applied Physics Laboratory, pp. 173–184 (May 28–29, 1975).

25. Grand, A. F. and C. D. Herrera. "Quantitative Measurement of Inhaled Hydrogen Cyanide," Southwest Research Institute, Internal Research, SwRI Project No. 01-9266 (January 1983).
26. Feldstein, M. and N. C. Klendshoj. "The Determination of Cyanide in Biological Materials," *J. Lab. Clin. Med.*, 55:166 (1954).
27. Caplan, Y. and R. Altman. "Microdetermination of Cyanide in Fire Fatalities," *28th Annual Meeting of American Academy of Forensic Sciences*, Washington, D.C. (1976).
28. Birky, M. M., B. M. Halpin, Y. H. Caplan, R. S. Fisher, J. M. McAllister and A. M. Dixon. "Fire Fatality Study," *Fire and Materials*, 3(4):211–217 (1979).
29. Wetherell, H. R. "The Occurrence of Cyanide in the Blood of Fire Victims," *J. Forensic Sci.*, 11(2):167–172 (1966).
30. Rieders, F., in I. Sunshine (ed.) *Methodology for Analytical Toxicology*, CRC Press, p. 115 (1971).
31. Reinke, R. W. and C. F. Reinhardt. "Fires, Toxicity and Plastics," *Modern Plastics*, pp. 94–98 (February 1973).
32. Brobeck, J. R., ed. *Best and Taylor's Physiological Basis of Medical Practice*. 9th ed., Williams and Wolkins, Baltimore, MD, Ch. 6, p. 52 (1973).
33. Barrow, C. S., Y. Alarie and M. F. Stock. "Sensory Irritation Evoked by the Thermal Decomposition Products of Plasticized Poly-Vinyl Chloride," *J. Fire and Materials*, 1:147–153 (1976).
34. Kaplan, H. L., A. Anzueto, W. G. Switzer and R. K. Hinderer. "Acute Respiratory Effects of Inhaled Polyvinyl Chloride (PVC) Smoke in Nonhuman Primates," *The Toxicologist*, Abstracts of Paper, Society of Toxicology, Incorporated, 26th Annual Meeting, 7(1):202 (February 1987).
*35. Kennah, H. E., M. F. Stock and Y. C. Alarie. "Toxicity of Thermal Decomposition Products from Composites," *J. Fire Sciences*, 5(1):3–16 (1987).
36. Anderson, R. C. and Y. Alarie. "Acute Lethal Effects of Polyvinylcholoride Thermal Decomposition Products in Normal and Cannulated Mice," *Abstracts of Papers, Society of Toxicology, Inc., 19th Annual Meeting*, Washington, D.C. (March 9–13, 1980).
37. Patra, A. L. "Comparative Anatomy of Mammalian Respiratory Tracts: The Nasopharyngeal Region and the Tracheobronchial Region," *J. Tox. Env. Health*, 17:163–174 (1986).
38. Patra, A. L., A. Gooya and M. G. Menache. "A Morphometric Comparison of the Nasopharyngeal Airway of Laboratory Animals and Humans," *The Anatomical Record*, 215:42–50 (1986).
39. Kaplan, H. L., A. Anzueto, W. G. Switzer and R. K. Hinderer. "Effects of Hydrogen Chloride on Respiratory Response and Pulmonary Function of the Baboon," *J. Tox. Env. Health* (in press).
*40. Hartzell, G. E., S. C. Packham, A. F. Grand and W. G. Switzer. "Modeling of Toxicological Effects of Fire Gases: III. Quantification of Post-Exposure Lethality of Rats from Exposure to HCl Atmospheres," *J. Fire Sciences*, 3 (May/June 1985).
*41. Hartzell, G. E., A. F. Grand and W. G. Switzer. "Modeling of Toxicological Effects of Fire Gases: VI. Further Studies on the Toxicity of Smoke Containing Hydrogen Chloride," *J. Fire Sciences*, 5(6):368–391 (November/December 1987).
*42. Hartzell, G. E., H. W. Stacy, W. G. Switzer and D. N. Priest. "Modeling of Toxicological Effects of Fire Gases: V. Mathematical Modeling of Intoxication of Rats by Combined Carbon Monoxide and Hydrogen Cyanide Atmospheres," *J. Fire Sciences*, 3(5):330–342 (September/October 1985).

43. Kaplan, H. L. (personal communication).

44. Levin, B. C., M. Paabo, J. L. Gurman, S. E. Harris and E. Braun. "Toxicological Interactions Between Carbon Monoxide and Carbon Dioxide," *Proceedings of 16th Conference of Toxicology*, Air Force Aerospace Medical Research Laboratory, Dayton, OH (October 1986).

45. Esposito, F. M. and Y. Alarie. "Inhalation Toxicity of Carbon Monoxide and Hydrogen Cyanide Gases Released During the Thermal Decomposition of Polymers," *J. Fire Sciences* (submitted).

46. Grand, A. F., H. L. Kaplan, J. J. Beitel, III, W. G. Switzer and G. E. Hartzell. "An Evaluation of Toxic Hazards from Full-Scale Furnished Room Fire Studies," *Fire Safety: Science and Engineering*, ASTM STP 882, T. Z. Harmathy, ed. American Society for Testing and Materials, Philadelphia, PA, pp. 330–353 (1985).

47. Kaplan, H. L., A. F. Grand and G. E. Hartzell. *Combustion Toxicology: Principles and Test Methods*, Lancaster, PA:Technomic Publishing Company, Inc. (1983).

NOTE: References noted with asterisk are contained in *Advances in Combustion Toxicology*.

BIOGRAPHY

Gordon E. Hartzell

Dr. Gordon E. Hartzell has been Director of the Department of Fire Technology at SwRI since 1978. His experience includes 20 years of industrial research and development in organic and polymer chemistry at the Dow Chemical Company. Before joining SwRI, he was a Research Associate Professor of Materials Science at the University of Utah.

Active in the field of fire technology since 1974, Dr. Hartzell has authored over 50 papers, largely in the areas of combustion toxicology and hazards to life due to fire. He is a member of ASTM Committee E-5 on Fire Standards and serves on working groups of International Organization for Standardization (ISO) TC92/SC3 which address toxic hazards in fire.

A member of the Society of Toxicology, Dr. Hartzell is a co-author of the book, *Combustion Toxicology: Principles and Test Methods*, Editor-in-Chief of the *Journal of Fire Sciences*, and editor of *Advances in Combustion Toxicology*.

Biological Studies of Combustion Atmospheres

D.A. PURSER

Huntingdon Research Centre
Huntingdon, Cambridgeshire, PEL8 6ES
England

W. D. WOOLLEY

Fire Research Station
Melrose Avenue
Borehamwood, Hertfordshire, WD6 2BL
England

ABSTRACT

There is evidence supported by statistical information from fire deaths that many fire fatalities occur as a result of incapacitation of the victims by the toxic products given off during the early stages of fires, thereby preventing escape from the fire, rather than from direct exposure to heat or other factors. As an essential part of understanding these problems, a study has been made of the mechanisms of incapacitation resulting from exposures to atmospheres of thermal decomposition products from polymeric materials. Under conditions approved by the Home Office Inspector, individual cynomolgus monkeys were exposed to atmospheres increasing in separate experiments from very low smoke concentrations until early signs of physiological effects were detected. Measurements were made of two kinds of physiological parameters: vital signs (respiration, electrocardiography and respiratory blood gases) and parameters indicating effects on the nervous system (electroencephalography, auditory evoked potentials, nerve conduction velocity). The atmospheres generated were designed to study the effects of hypoxia, hypercapnia, carbon monoxide, hydrogen cyanide and thermal decomposition products from wood, polyacrylonitrile, polyurethane foam, polypropylene, polystyrene and nylon produced under pyrolytic or oxidative conditions at a range of temperatures. The main findings were that the

© Crown Copyright 1982
Text of a lecture given at the conference 'Smoke and toxic gases from burning plastics' 6-7 January 1982 organised jointly by QMC Industrial Research Ltd. and the Fire Research Station.

composition and hence the toxicity of the products from individual materials could vary considerably depending upon the different conditions of temperature and degree of oxygenation under which they were decomposed. However, despite the great complexity in chemical composition of the test atmospheres, the basic toxic effects on the animals were relatively simple, and for each individual atmosphere the toxicity was always dominated by one of these factors; carbon monoxide, hydrogen cyanide, or irritants. The role of each of these factors in causing incapacitation in real fires is discussed.

INTRODUCTION

IN ORDER TO understand the toxic hazards presented by polymeric materials in fires a number of different approaches have to be used. One, as Dr. Woolley has described [1] is to examine the physical and chemical properties of real fires and of both large and small scale fire models, and to attempt to predict the likely consequences to human fire victims by extrapolation from existing toxicity data on individual fire gases. The second approach, outlined by Professor Harland [2] is to examine actual fire victims and to try to work back to the situation which is likely to have been responsible for their condition. The third approach is to examine the biological effects of combustion products in the laboratory, and this is a vital link between the chemical and pathological approaches in providing meaningful information by direct physiological observations which has therefore necessitated the use of an animal model.

Much of the information to date which has been obtained from bioassay studies, and which has emerged from pathology of fire victims, is relevant in understanding the direct causes of death resulting from exposure to fire atmospheres. The concept of physiological incapacitation following sublethal exposures to the toxic species of combustion products has been known for some time but it is only in recent years that the importance of effects of this kind has been recognised in limiting escape so that persons remain close to the fire. Their death subsequently being attributed to carbon monoxide poisoning and other well established factors (e.g. oxygen depletion or hot gases).

There is now considerable supporting evidence from studies on fire death statistics [3,4,5] that the combustion products given off during fires are an important cause of fire deaths, and that the actual lethality of fire products may be less important than their ability to incapacitate the victim with a large percentage (61% for dwellings in 1979) [5] of fatal casualties being found in the room of origin of the fire.

Studies of thermal decomposition and combustion products of a wide range of materials under both large and small scale fire conditions [6,7] have shown them to contain a variety of potentially toxic chemical species. Some, such as carbon monoxide and hydrogen cyanide, are narcotic and can incapacitate occupants in a fire by affecting the nervous and cardiovascular systems, while others are irritant and incapacitate by effects upon the eyes and upper and lower respiratory tracts causing respiratory distress and behavioural incapacitation.

In understanding the complex interactions which may take place between the fire atmosphere and a building occupant it is essential to have a detailed knowledge of the chemical and physical properties of the atmosphere. This information has been reported by Woolley [1], derived from studies over a number of years designed to characterise particularly the chemical nature and obscurational effects of the smoke and gases from model and full scale experimental fires involving a range of common polymers.

This work has shown that fire atmospheres are not only hot and depleted in oxygen but are complex mixtures of chemical products which contain toxic and irritant species and may be rich in solid particulate and liquid aerosol matter. He has also shown that the many hundreds of chemical components may change qualitatively and quantitatively with different fire conditions, being markedly dependent upon the oxygen availability and temperature. Also of particular importance is the type of fire and the possible exposure pattern. For example under smouldering (i.e. non-flaming) conditions as in the case of a smouldering bed or chair, then there may be a very long period (many hours) during which there is a gradual increase in the local concentrations of smoke and gases with the victim inhaling the products for some time before there is an awareness of a problem. This may be especially significant to persons asleep, infirm or intoxicated. In other conditions with flaming types of fire there may be more immediate hazards to life where persons may be confronted suddenly with hot, oxygen depleted and toxic environments sufficient to bring about unconsciousness and death in very short periods of time.

The most extensively studied and probably the most important product evolved during fires is carbon monoxide, which has been shown to be present at fatal levels in 50% of the fatalities studied in Professor Harland's work [8]. Hydrogen cyanide is also known to be present in some fire atmospheres, and many are known to contain highly irritant compounds such as hydrogen chloride and oxygenated organic species (e.g. acrolein and formaldehyde).

Much less is known as to possible effects when several narcotic gases are present together. The contribution of hydrogen cyanide to fire deaths for instance has been difficult to assess because it is usually accompanied by carbon monoxide, and similar problems are experienced when considering narcotic gases in combination with irritants. Most fire atmospheres contain a daunting array of different organic products and, even when the decomposition and/or burning of only a single polymer is involved, some products will be formed whose toxicity is completely unknown. Fortunately to date known cases of 'supertoxicants' are extremely rare but compounds of this type such as 'TMPP' [9] have caused concern in the past. Even in cases where toxicity data are available on individual products the data are seldom in the form where it is possible to predict time to incapacitation or effects on escape capability.

The work described in this report was carried out to fill a vital gap in knowledge about the sublethal effects of the decomposition and combustion products of polymeric materials as relevant to the possible reduction in escape potential of persons exposed to fire. The work has involved the development of a suitable model to study, and hence describe, the incapacitative effects of fire atmospheres. This type of work can only be carried out with biological procedures and is essential in understanding the mechanisms which lead to the loss of many hundreds of lives each year in fires in the UK alone, and many thousands throughout the world.

The approach has been to carry out a limited number of carefully selected exposures at low concentrations of toxicant(s) to detect the first signs of incapacitation, with a battery of complex measurements of vital signs and neurological functions to ensure that the maximum amount of information was available from each exposure. We have avoided statistical 'dose response' approaches in much of the work which would have involved large numbers of experiments and we have concentrated on selected (and meaningful) exposures to the products from a range of the common polymers which are confronting humans (persons escaping or involved in fire, Fire Brigades and fire researchers) causing injury and death with some regularity.

The overall work has led to a clearer understanding of the mechanisms of incapacitation but the method of approach differs from that described by Dr. Klimisch [10] in his paper, in that we have not attempted to produce quantative indices of toxicity for specific materials. We must therefore emphasise that the results for individual materials (obtained from limited number of exposures) should not be used in isolation but form part of the overall pattern of likely incapacitative effects.

The model consisted of cynomolgus monkeys exposed individually under conditions approved by the appropriate Home Office Inspector, to test atmospheres consisting of sublethal levels of individual gases, or thermal decomposition products of polymeric materials (polyacrylonitrile, flexible and rigid polyurethane foam, wood, polypropylene, polystyrene and nylon) produced under pyrolytic or oxidative conditions at a range of different temperatures. It must be emphasised that this work has not been carried out to develop a routine test procedure, though tests of this kind are being proposed in other countries and use laboratory animals for the toxicological assessments. Points which have emerged from this present research relevant to test procedure, together with the reservations of the authors, will be given.

EXPERIMENTAL REGIME

The aim of each experiment was to produce atmospheres for each material which would produce initial signs of incapacitation in the animals within a 15-30 minute period. This was achieved by carrying out studies in which the animals were exposed to atmosphere concentrations increasing on separate occasions from very low concentrations to a concentration where early signs of a physiological response were seen. Each of 3 animals was then exposed to this selected atmosphere concentration and the physiological responses were measured.

Atmospheres were maintained in a dynamic steady state composition using a modified version, with modified operating conditions, of the Kimmerle system (DIN 53436) [7] by introducing the test material into a tube furnace at a constant rate and diluting the products with air. Throughout the experimental session a battery of physiological parameters was measured and the clinical condition of each animal was monitored closely. The measurements were of two kinds; firstly of vital signs—respiration, electrocardiography (ECG) and blood levels of respiratory and toxic gases, and secondly special tests of neurological and neuromuscular function—electroencephalography (EEG), of the cortical evoked response to auditory stimuli, and peripheral nerve conduction velocity.

The results reported here represent a summary of a series of 3 studies, each carried out on a separate set of animals.

The initial series of experiments was carried out on gas mixtures of the general type likely to be encountered in fires and to test the suitability of the animal model. Exposures were as follows:

Hypoxia (10% and 15% oxygen)	balance nitrogen
Hypercapnia (5% carbon dioxide)	balance air
Carbon monoxide 1000 ppm	balance air
Hydrogen cyanide 60 ppm	balance air

The experiments on carbon monoxide and hydrogen cyanide were later extended to cover the range of concentrations encountered in the smoke experiments. CO (1000–8000 ppm) (unpublished), HCN 80-150 ppm [11].

For the second set of experiments animals were exposed to atmospherers produced when each of 3 materials was pyrolysed at each of 3 temperatures and diluted with air. These were chosen to cover as nearly as possible the full temperature range known to occur in fires, and where oxygen access to the polymer might, for a number of reasons, be restricted.

Polyacrylonitrile)	
)	
Flexible polyurethane foam)	pyrolysed at 900, 600 and 300 °C
)	
Scots pine wood)	

For the third set of experiments the range of polymers was extended and the effects of non-flaming and flaming oxidative conditions on test atmosphere toxicity were examined at middle range temperatures (440–700 °C). These were chosen to embrace the conditions which are known to give the greatest complexity of products, particularly with polymers sensitive to oxidation in product formation. Each animal was exposed as follows:

Rigid polyurethane foam:	non-flaming oxidative	600 °C
Polypropylene:	pyrolytic	600 °C
	non-flaming oxidative	450 °C
	flaming oxidative	700 °C*
Polystyrene:	pyrolytic	600 °C*
	non-flaming oxidative	500 °C
Nylon:	non-flaming oxidative	440 °C

(*Single exposures)

The polymers used for these experiments were chosen as being typical of those currently found within buildings. The various decomposition conditions used for each polymer were chosen to ensure that each experiment provided the maximum amount of relevant information.

EXPERIMENTAL METHODS

Cynomolgus monkeys (Macaca fascicularis, body weight 3–4 kg) were exposed individually as shown in Figure 1. Each test session consisted of 3 consecutive 30 minute periods:

1. Pre-exposure period during which baseline levels of physiological parameters were established.
2. Exposure period during which the effects of the test atmospheres were examined.
3. Recovery period during which the parameters under examination were observed as they returned towards pre-exposure levels.

MATERIALS

The wood used was Pinus sylvestris, as a typical softwood. The polyacrylonitrile was used in fibre form (98%). The flexible polyurethane foam was a typical polyether form with a toluene diisocyanate (TDI) base, and the rigid polyurethane foam was also a polyether type with a diphenylmethane diisocyanate (MDI) base. The polypropylene was a commercially available type (isotactic) as were the polystyrene and nylon samples which were used in a granular form.

Atmosphere Generation and Physiological Methods

The apparatus used to generate the test atmosphere is shown in Figure 1. The sample, in a boat, was placed in the pyrolysis tube at the

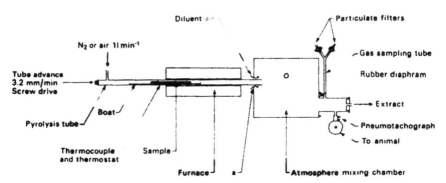

Figure 1. Atmosphere generation system.

entrance to the tube furnace. The pyrolysis tube was then advanced through the furnace at a constant rate (3.2 mm/min) so that the sample was gradually carried through the hot zone of the furnace, a nitrogen or air flow of 1 litre/min being passed through the pyrolysis tube to provide either pyrolytic or oxidative conditions, and to carry the products into the atmosphere mixing chamber. The test atmosphere was then produced in the atmosphere mixing chamber by mixing the products with diluent air, drawn in under subatmospheric pressure through a gap between the entry port and the outer surface of the pyrolysis tube at Point (a). The mixed atmosphere was withdrawn through the extension piece for inhalation via a pneumotachograph (for measuring the air flow resulting from the animal's respiration).

Measurements were made in the atmosphere mixing chamber of temperature, the optical density of the smoke, the particulate concentrations (from samples deposited on a filter), carbon monoxide, carbon dioxide, oxygen and hydrogen cyanide. The test atmosphere was characterised in terms of nominal atmosphere concentrations (nominal atmosphere concentration = mass of material introduced into the furnace each minute divided by the diluent air flow rate). On two occasions 'grab' samples of polyacrylonitrile atmospheres were subjected to a gas chromatographic/mass spectrometric 'finger print' analysis. Respiratory tidal flow, tidal volume, respiratory rate and respiratory minute volume (RMV) were measured continuously. At the same time recordings were made of the electrocardiogram (ECG) and electroencephalogram (EEG) (from electrodes implanted on the surface of the cerebral cortex). The power spectrum of the EEG was filtered into each of the four major wave bands (delta 1–4 Hz, theta 4–8 Hz, alpha 8–13 Hz, beta 13–30 Hz), and the distribution of the EEG activity was measured in each successive 30 second period throughout the test session by integrating the activity in each wave band. In general a preponderance of delta and theta activity is indicative of a drowsy or unconscious state and is characteristic of central nervous system (CNS) depression. The presence of desynchronised high frequency beta activity is characteristic of an alert, actively thinking state, and alpha activity represents a relaxed but waking condition. When the total amount of activity is low, and the predominant activity is in the delta band the subject is usually approaching a comatose state. During some exposures to single gases, measurements were made of the motor nerve conduction velocity of the ulnar nerve, and also of click-evoked auditory cortical responses. On some occasions measurements were made of arterial blood pressure, PO_2, PCO_2, pH and also blood carboxyhaemoglobin levels (COHb) or blood cyanide.

RESULTS

Experiments with Single Gases

The results of the individual gas exposures confirmed the suitability of the model in that all the physiological parameters measured were affected, although in most cases the effects were more or less at a threshold level. Exposures to 10% oxygen resulted in an EEG picture of CNS depression and almost immediate effects on nerve conduction velocity and the auditory evoked potential, together with some hyperventilation and occasional ECG abnormalities.

Exposure to carbon monoxide at 1000 ppm produced COHb levels of 30% by the end of the 30 minute exposure. This was accompanied by reductions of nerve conduction velocity and in some cases the animals showed severe CNS depression. HCN at 60 ppm produced a slight CNS depression and a slight hyperventilation. At higher levels (80-150 ppm) the hyperventilation was marked and followed by a severe CNS depression and cardiac abnormalities. Exposures to CO_2 had little effect beyond causing a threefold increase in RMV and evidence of slight EEG depression at the end of the exposure.

Results of Experiments with Polymeric Materials

After studying the incapacitating effects of 16 different test atmospheres produced from seven different materials decomposed under a range of temperatures and degrees of oxygenation, two major points have emerged which are of special interest:

1. Variations in temperature and degree of oxygenation were shown to have a profound influence upon the composition and hence the types of toxic effects of the decomposition and combustion products from individual materials.
2. In spite of the great complexity in chemical composition of the test atmospheres the basic toxic effects on the animals were relatively simple, and for each individual atmosphere the toxicity was always dominated by the narcotic effects of either carbon monoxide or of

hydrogen cyanide or by the effects of combinations of irritant chemical species.

The Effects of Nitrogen Containing Polymers and Wood Pyrolysed at 3 Temperatures

The effects of temperature on thermal decomposition products can be shown from the results of the series of experiments which involves the pyrolysis of nitrogen containing polymers and wood (Table 1).

Generally speaking at temperatures above 700° C many decomposition products are unstable and relatively simple atmospheres of low molecular weight products are formed, whereas at lower temperatures larger molecular fragments and denser smokes are formed [6,12]. Although this was true of the three materials examined, the patterns of thermal decomposition of the two nitrogen containing polymers, polyacrylonitrile and flexible polyurethane foam showed considerable differences. For polyacrylonitrile the atmospheres produced were basically similar at all 3 temperatures, with relatively low smoke and particulate concentrations, the main toxic species being HCN, although small quantities (6-40 ppm) of acetonitrile, benzonitrile and acrylonitrile were also detected. The main effect of increasing the temperature was to increase the yield of HCN from 1.7% by mass of original polymer at 300° C to 4.5% at 600° C and 13.9% at 900° C, and to decrease slightly the concentration of smoke and particulates, so that at 900° C the atmosphere was almost clear. The *pattern* of toxicity was very similar for each pyrolysis temperature (although the *effects* were more severe at higher temperatures), and was almost identical to that produced by an equivalent concentration of HCN gas alone. The only difference between different temperatures was a slight smoke at 300° C and 600° C which was accompanied by a slight background hyperventilation in the animals.

For flexible polyurethane foam there were considerable differences between the atmospheres produced by pyrolysis at different temperatures. At 900° C the atmosphere was clear, with high yields of CO (38.6%) and HCN (5.3%), and low levels of smoke and particulates. The toxicity pattern was very similar to that produced by HCN gas or polyacrylonitrile products, which agrees well with the chemical analyses of products by Woolley and his colleagues [12,13], who showed that the high temperature chromatograms produced for polyurethane foam are very similar in qualitative terms to those for polyacrylonitrile, consisting of HCN and other nitriles produced from the breakdown of toluene diisocyanate (TDI) and various hydrocarbon fragments from the breakdown of the polyol. At lower temperatures, however, the products are very different. At 300° C flexible polyurethane foam depolymerises into TDI and the polyol. The TDI is not released directly, but undergoes a type of polymerisation to form a dense yellow particulate smoke, and up to 1% by mass of original foam may be released as TDI monomer,

Table 1. Composition of test atmospheres and clinical signs.

Test atmosphere		Nominal atmosphere concentration (mg/l)	Particulate concentration (mg/l)	Smoke OD/m	CO (ppm)	CO₂ (ppm)	HCN (ppm)	Time to incapacitation (min)	Clinical signs
Polyacrylonitrile pyrol	900°C	1.43	0.07	0.08	<20	—	117	8	Hypervent. then semicon.
Polyacrylonitrile pyrol	600°C	3.99	0.45	0.71	<10	—	159	14	Hypervent. then semicon.
Polyacrylonitrile pyrol	300°C	8.20	0.44	0.53	<10	—	126	20	Hypervent. then semicon.
Flex. Pu pyrol	900°C	2.77	0.12	0.05	917	—	130	14	Hypervent. then semicon.
Flex. Pu foam pyrol	600°C*	9.55	1.15	2.12	1400	—	10	im	Dysp. and hypervent.
Flex. Pu foam pyrol	300°C	16.05	1.30	1.55	<10	—	<10	im	Dysp. and hypervent.
Wood (Scots Pine) pyrol	900°C	9.60	0.14	0.17	1902	1133	—	25	No hypervent. semicon.
Wood pyrol	600°C	14.60	0.75	1.01	1488	1167	—	25	Dysp. hypervent semicon.
Wood pyrol	300°C	90.50	3.73	0.20	1767	3400	—	26	Dysp. hypervent. semicon.
Rigid Pu foam N.F. Ox.	600°C	1.83	0.03	0.00	1187	2467	108	23	Hypervent. then semicon.
Polypropylene F. Ox.	700°C**	11.5	0.07	<1.00	1700	7000	—	17	Dysp. then relaxed, semicon.
Polypropylene pyrol.	600°C	17.2	1.45	3.2	22	<500	—		No effect
Polypropylene NF. Ox.	450°C	7.4	0.98	2.7	1267	3633	—	im	Dysp. and hypervent.
Polystyrene pyrol	600°C**	7.3	1.00	3.6	<10	<500	—	im	Dysp. and hypervent.
Polystyrene NF. Ox.	500°C	7.3	1.08	2.7	430	1550	—	im	Dysp. and hypervent.
Nylon. NF. Ox.	440°C	4.7	1.70	2.9	137	933	39	im	Dysp. and hypervent.

All figures mean of 3 exposures except * mean of 2 exposures; ** single exposures.
Abbreviations: Pu = polyurethane, Pyrol = pyrolysed, F = flaming, NF = non-flaming, Ox = oxidative, Hypervent. = hyperventilation, Dysp. = dyspnoea, Semicon. = semiconscious, — = not measured, im = some degree of immediate incapacitation.

with the polyol remaining in the furnace as a brown liquid. At 600° C the products are similar but some breakdown of the smoke and polyol gives rise to HCN and CO, and there is little or no TDI monomer. At both of these temperatures the atmospheres produced were highly irritant to the animals.

As with the polyacrylonitrile, wood smoke atmospheres were fairly similar at the 3 pyrolysis temperatures. At 900° C a clear atmosphere was produced, with a high yield of CO (23.1%) and at lower temperatures more smoke and particulates were produced and less CO.

In all experiments the temperature in the atmosphere mixing chamber, from which smoke was supplied for inhalation studies, was between 26 and 29° C, only slightly above room temperature.

Toxicity

From the toxic signs observed during the exposures, the test atmospheres could be classified into 3 groups:

Group 1   Atmospheres containing narcotic and potentially toxic concentrations of CO:
          Wood pyrolysed at 900° C, 600° C and 300° C.
Group 2   Atmospheres containing narcotic and potentially toxic concentrations of HCN:
          Polyacrylonitrile pyrolised at 900° C, 600° C and 300° C.
          Flexible polyurethane foam pyrolysed at 900° C.
Group 3   Irritant atmospheres containing high concentrations of particulates and low concentrations of CO and HCN:
          Flexible polyurethane foam pyrolysed at 600° C and 300° C.

Group 1—Atmospheres Containing Narcotic and Potentially Toxic Concentrations of CO

The physiological signs produced by the Group 1 atmospheres were all similar and the most significant toxic agent common to them all was CO. The effects on the physiological parameters are shown in Figure 2. The onset of intoxication was insidious, and the first signs of toxicity always occurred at the end of the exposure period (at 20–30 minutes) when the animals had venous COHb levels approaching 40% for smoke atmospheres containing 1100–2000 ppm CO. The animals were generally relaxed throughout the exposures, and there was often a slight decrease in respiration, especially with pure gas atmospheres and the clear atmospheres produced by pyrolysis at 900 °C. At the point during the exposure when incapacitation occurred, its onset was rapid, and was often preceded by a brief period of activity by the animal. It consisted of a state of semiconsciousness accompanied by a slow, irregular heart beat and slow-wave EEG activity (Figure 2). The condition of the animals im-

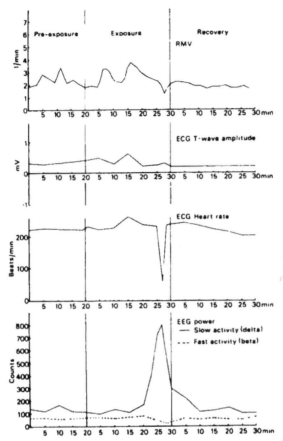

Figure 2. Physiological effects of an exposure to an atmosphere containing CO (1850 ppm) — wood pyrolysed at 900°C Experiment C18 (821).

proved rapidly during the first 5 minutes after they were given air to breathe, but a reduced respiration continued throughout the recovery period. Unlike the atmospheres containing HCN, small changes in atmosphere concentration (and thus CO concentration) did not obviously affect the time to incapacitation, but incapacitation was more likely with CO concentrations approaching 2000 ppm. In fact the occurrence of incapacitation was somewhat unpredictable and for CO levels of between 1000 and 1500 ppm and COHb levels of from 30-40%, the collapse from an apparently normal state into one of semiconsciousness often failed to occur, particularly with inactive animals.

For the lower temperature wood smoke atmospheres, particularly at 300° C, there were some signs of irritance during the early stages of the exposure (see Group 3 below).

Group 2—Atmospheres Containing Narcotic and Potentially Toxic Concentrations of Hydrogen Cyanide

The dramatic signs of incapacitation produced by atmospheres containing HCN (Figure 3) began at some time during the exposure period indicated by a short episode of a marked increase in respiration (hyperventilation), the RMV increasing by up to four times, after 1-5 minutes of which the animals became semiconscious. This was accompanied by EEG signs of severe cerebral depression, loss of muscle tone and marked effects upon the heart and circulation, including a significant decrease in heart rate, arrhythmias and changes in the ECG waveform. This hyperventilatory episode was almost certainly caused by the stimulatory effects of cyanide upon respiration (via the carotid body chemoreceptors), which since it was taken in by the inhalation route, resulted in a positive feedback situation, the inhaled HCN causing hyperventilation which increased the rate of HCN uptake and in turn

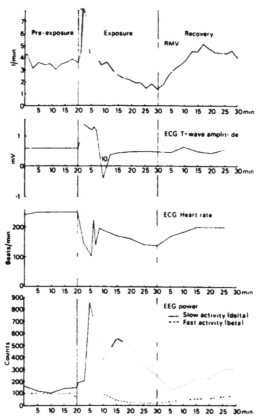

Figure 3. *Physiological effects of an exposure to an atmosphere containing HCN (163 ppm)—polyacrylonitrile pyrolysed at 600°C Experiment C30 (825).*

provided a stronger hyperventilatory stimulus. The threshold for this phenomenon was at approximately 80 ppm HCN, and the time at which it occurred during the exposure period was proportional to the chamber HCN concentration, (correlation: $R = 0.8$, $P < 0.001$) varying from 20 minutes at 90 ppm to an immediate response at 200 ppm HCN, and producing blood cyanide levels of 78-180 μMoles/l after a 30 minute exposure. The effects of the cyanide containing polyacrylonitrile and flexible polyurethane smoke atmospheres were almost identical to those produced by atmospheres containing similar concentrations of HCN gas alone [11].

Immediately after the hyperventilatory episode the condition of the animals appeared to be at its worst in some ways, particularly with regard to the effects upon the cardiovascular system. However, during the next 5 minutes the condition of the animals rather surprisingly improved. The heart rate and rhythm improved and the respiration rate and volume decreased, the animal occasionally regaining consciousness for periods of a few seconds. The most likely explanation for this is that the depressive effect of cyanide on the respiratory centre in the medulla [14] caused the hyperventilation to subside, which in turn resulted in a decreased rate of HCN uptake enabling the animal to recover slightly. From this point the animals went into a slow decline over the remainder of the exposure period, the rate and volume continuing to decrease until a pattern of slow deep breaths occurred with pauses at the end of each expiration, a pattern typical of anaesthetised animals, which eventually led to a cessation of breathing in some cases, and would have been fatal if the exposure had not been terminated.

The important point about this response pattern is that although HCN is capable of producing a rapid incapacitation, the animal can remain in this semiconscious state for some time, and yet recover almost completely within 5-10 minutes of the end of the exposure. When one animal was exposed to 196 ppm HCN in a polyacrylonitrile atmosphere it became unconscious after 2 minutes, but made a rapid recovery after the end of the 30 minute exposure period.

The only difference in response between the atmospheres was that whereas with the clear, high temperature, atmospheres the animals showed no response until the hyperventilatory episode began, with the lower temperature atmospheres which contained more smoke and particulates (polyacrylonitrile pyrolysed at 600° C and 300° C) there was a tendency for a slight background hyperventilation to occur until the animal became semiconscious.

Group 3—Irritant Atmospheres Containing High Concentrations of Particulates and Low Concentrations of CO and HCN

The signs produced by Group 3 atmospheres appeared to be related to the high smoke and particulate concentrations. There were no obvious signs of intoxication but marked hyperventilation and respiratory distress occurred. The initial response consisted of a disturbed pattern of respiratory (dyspnoea) during the first 1-3 minutes with rapid inspirations punctuated by pauses in respiration, often with decreases in both respiratory rate and tidal volume. This gradually gave way to a marked hyperventilation consisting of increases in both respiratory rate and tidal volume and a more regular breathing pattern, although occasional sudden inspirations and pauses occurred. The hyperventilation gradually decreased towards the end of the exposure period and the respiration pattern was more normal during the recovery period, although there were occasional periods of dyspnoea. The severity of both the initial dyspnoea and the later hyperventilation depended to some extent upon the nominal atmosphere concentration. The animals appeared normal as soon as the exposure was terminated, but with flexible polyurethane foam pyrolysed at 600° C signs of an acute irritant lung reaction, with dyspnoea, pulmonary oedema and alveolar congestion, occurred 12-24 hours after exposure. No such late effects were observed following exposure to atmospheres from polyurethane foam pyrolysed at 300° C.

For the lower temperature wood smoke atmospheres, particularly at 300° C transient signs of irritance were observed consisting of an initial dyspnoea followed by a slight hyperventilation during the first 5 minutes of exposure (Figure 4).

The Effects of Oxidative Decomposition on Polymer Toxicity

All non-flaming decomposition exposures were carried out at the highest furnace temperature at which stable non-flaming conditions could be maintained. For the single exposure to flaming oxidative conditions (polypropylene) stable conditions were achieved at a furnace tem-

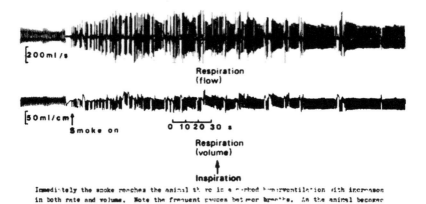

Immediately the smoke reaches the animal there is a marked hyperventilation with increases in both rate and volume. Note the frequent pauses between breaths. As the animal becomes used to the atmosphere the pattern becomes more regular but the hyperventilation is maintained.

Figure 4. The effects upon respiration of exposure to a particulate rich atmosphere Experiment C22 (821) Wood 300°C.

perature of 700° C. Two materials produced clear atmospheres, rigid polyurethane foam under non-flaming conditions and polypropylene under flaming oxidative conditions, all other atmospheres were dominated by smoke and particulates.

Although the range of polymers and methods of decomposition were different from those of the previous experiments, it was possible to classify the test atmospheres into the same three basic groups on the basis of their composition and the toxic signs observed.

Group 1 Atmospheres containing narcotic and potentially toxic concentrations of CO:
Polypropylene flaming oxidative decomposition at 700° C.

Group 2 Atmospheres containing narcotic and potentially toxic concentrations of HCN:
Rigid polyurethane foam non-flaming oxidative decomposition at 600° C.

Group 3 Irritant atmospheres containing high concentrations of particulates and low concentrations of CO and HCN:
Polypropylene non-flaming oxidative decomposition at 450 °C
Polypropylene pyrolytic decomposition   at 600 °C
Polystyrene non-flaming oxidative decomposition   at 500 °C
Polystyrene pyrolytic decomposition   at 600 °C
Nylon non-flaming oxidative decomposition   at 440 °C

The atmosphere produced by the non-flaming oxidative decomposition of rigid polyurethane foam (based on diphenylmethane diisocyanate —MDI) and its toxic effects were very similar to those produced by the pyrolytic decomposition of flexible polyurethane foam at 900° C, there being very little particulate matter and no smoke, unlike the dense atmosphere produced by the flexible foam at 600° C under pyrolytic conditions. Two factors which may explain this are that the yield of HCN from polyurethane foam is greater at mid-range temperatures under oxidative conditions than under pyrolytic conditions [6], and that MDI-based foams do not depolymerise and form isocyanate smokes as do the flexible foams but remain in the furnace to be decomposed into smaller fragments [15]. The main obvious toxic products were CO (yield 75.7%)* and HCN (yield 6.6%). There was also a 209% conversion to CO_2.

The effects of degrees of oxygenation on the decomposition products of hydrocarbon polymers can be demonstrated by the results obtained from polypropylene. Under pyrolytic conditions at 600° C the atmosphere proved innocuous to the animals, with no signs of toxicity or physiological response, although a high nominal atmosphere concentration of 17.2 mg/litre was used and the atmosphere was rich in smoke and particulates. Under non-flaming conditions at 450° C a particulate and

---

*Figures in parentheses indicate percentage by mass of original polymer.

smoke rich atmosphere was also produced which was, however, extremely irritant, the signs being very similar to those produced by the pyrolysis of flexible polyurethane foam at 300° C and 600° C, consisting of a transient dyspnoea with breath holding and bradycardia, followed by a prolonged hyperventilation throughout the remainder of the exposure. As with the exposures to flexible polyurethane foam at 600° C, the animals showed signs of an acute inflammatory lung reaction 12–24 hours after exposure, with coughing and nasal and buccal mucus discharge, which lasted for up to 48 hours, and in one animal gave rise to pulmonary oedema and alveolar congestion.

The reasons for the difference between the pyrolytic and oxidative atmospheres can be understood by reference to Table 2, which shows the products obtained by Woolley, Ames and Fardell [12] at 500° C. Under pyrolytic conditions the products consist of hydrocarbon fragments, which are not considered to be irritant, while under non-flaming oxidative conditions an additional range of products if formed, some of which are known to be highly irritant. However, under flaming oxidative conditions at 700° C most of these products are consumed, and a clear atmosphere was produced with high yields of CO (17.3%) and CO_2 (105.5%). The toxic effects were those of CO poisoning and were very similar to those produced by the pyrolysis of wood.

The atmosphere produced by the non-flaming oxidative decomposition of polystyrene was also somewhat irritant and contained small amounts of CO, producing irritant effects similar to those of polypropylene, but less severe. However, with this material the atmosphere produced under pyrolytic conditions was also moderately irritant, possibly due to the presence of styrene monomer. An irritant smoke was also produced when nylon was decomposed under non-flaming oxidative conditions.

Significance of Results

In addition to the observations about the intoxication by CO or HCN, and the irritancy of some smoke components, the work has also shown the somewhat unexpected finding that for every atmosphere studied, the toxicity picture was dominated by *one* of these influences, and that in spite of the fact that they could occur in combination, or that a complex mixture containing many other products may be present, the toxicity patterns were relatively simple and almost indistinguishable from those produced by the individual gases CO or HCN at the same concentration, or by irritants.

At temperatures of 700° C or above, when many decomposition products are unstable, relatively simple atmospheres of low molecular weight products are formed. Thus at 900° C polyacrylonitrile, polyurethane foam and wood produced clear atmospheres, the former consisting mainly of HCN with small amounts of other nitriles, and the latter pro-

Table 2. Pyrolytic and oxidative decomposition products of polypropylene at 500°C showing % yields of major irritants and Threshold Limit Values (TLVs) where available from American Conference of Governmental Industrial Hygienists 1980.

MS interpretation	Pyrolysis yield[b] (%) ($\times 10^{-1}$)	Oxidation yield[b] (%) ($\times 10^{-1}$)	TLV (ppm)
Ethylene	10.4	8.1	
Ethane	3.7	2.1	
Propene	18.6	18.4	
Cyclopropane	0.5	0.3	
Formaldehyde [a]	—	33.2	2
Propyne	0.2	—	
Acetaldehyde [a]	—	35.0	100
Butene	9.6	20.1	
Cyclobutene	0.3	0.8	
Methyl vinyl ether [a]	—	10.4	
Acetone [a]	—	38.4	750
Butane	1.2	—	
Methyl propane	0.4	—	
Methyl butane	4.0	—	
Butenone [a]	—	1.3	
Methyl butene	29.7	12.9	
Pentanol [a]	—	12.5	
Cyclopentane	0.5	1.4	
Pentadiene	1.3	—	
Crotonaldehyde [a]	—	7.7	2
Ethylcyclopropane	0.1	—	
Methyl vinyl ketone [a]	—	2.8	
Methyl ethyl ketone [a]	—	4.7	200
Hexane	0.9	1.2	
Cyclohexane	32.2	19.3	
Hexadiene	3.7	2.2	
Hexyne	—	1.3	
Benzene	6.7	5.1	
Methyl propyl ketone [a]	—	1.9	
Pent-2-ene-4-one [a]	—	7.5	
Phenol [a]	—	11.6	5
Toluene	2.4	16.1	
Methyl cyclohexadiene	2.1	0.1	
Xylene	6.1	0.2	
Styrene	5.6	4.0	

a Oxygen-containing products (underlined).
b Weight percent conversion of polymer.
from: Woolley, Ames and Fardell (1979).

ducing mainly carbon monoxide. It is perhaps not surprising that for atmospheres such as these the toxicity patterns were almost indistinguishable from those produced by the individual gases at the same concentrations. This is not to say that a contribution to the toxicity from the other atmosphere constituents can be eliminated, indeed there were indications of some rather subtle and unexpected effects from products other than HCN in atmospheres produced by polyacrylonitrile pyrolysed at 900° C [11] but it is extremely unlikely that any other substance made a major contribution. For some irritant atmospheres, such as those produced by flexible polyurethane foam at 300° C the effects may also be due to a single product, while for others such as the atmospheres produced by the non-flaming oxidative decomposition of polypropylene at 450°C, a number of known irritant chemical species can be identified, but the basic effects of all irritant atmospheres were found to be similar.

However, some atmospheres, particularly those produced at middle range temperatures, contain many different chemical species, including various combinations of HCN, CO and irritants. In these instances a complex pattern of toxicity might be expected, with each component making a contribution to the overall toxicity, and it is true that sometimes there was a likelihood of, and some evidence for both slight additive and synergistic effects. However, the surprising outcome was that for every case so far examined it can be said that the toxicity picture was dominated by only one factor, either CO, or HCN or irritance. For instance the atmosphere produced by the high temperature pyrolysis of flexible polyurethane foam was very similar to that produced by the pyrolysis of polyacrylonitrile, except that the former contained toxic concentrations of CO (800-1100 ppm) in addition to toxic concentrations of HCN (85-141 ppm) [16]. If there was a significant additive or synergistic effect of these two gases on time to incapacitation, it might be expected that the polyurethane foam atmospheres would produce incapacitation much sooner than the polyacrylonitrile atmospheres for a given HCN concentration.

In practice there was very little difference between the two atmospheres, the signs produced by both being identical to those produced by HCN gas alone, and for each class of atmosphere there was a statistically significant relationship between HCN concentration and time of incapacitation as follows:

	R	P
HCN gas	0.96	0.02-0.05
Polyacrylonitrile products	0.89	<0.001
Polyurethane foam products[ø]	0.85	0.01-0.02

ø both rigid and flexible all atmospheres

and although on average the polyurethane foam atmospheres did appear to produce incapacitation slightly earlier than did the polyacrylonitrile atmospheres (P = 0.01-0.02) the difference in time to incapacitation was only 10 percent.

When polyurethane foam was pyrolysed at 600° C, a highly irritant particulate rich atmosphere was produced which contained toxic concentrations of CO but the main effects of exposure were signs of irritance. In contrast to this, the atmosphere produced by the pyrolysis of wood at 600° C contained smoke and particulates, and some transient signs of irritance were observed early in the exposure, but the dominant effects of the 30 minute exposure were undoubtedly of CO poisoning and were very similar to those produced by the gas alone.

It is conceivable that greater degrees of interaction between the three main toxic factors may be possible if combination atmospheres are generated where the concentrations of each factor are adjusted to promote such effects, although serious interactions between the main narcotic gases CO and HCN are unlikely since both we and other investigators have found only minor interactions [16,17,18] or none [19] over a range of combination concentrations.

With regard to irritants the situation is more complex, as it is quite possible that considerable interactions occur between the effects of individual irritant products. However, although we have seen both irritant effects and narcotic CO effects in some atmospheres, irritance being evident during the initial stages of the exposure and CO induced narcosis towards the end, both we and Higgins et al [17] have seen no evidence that the presence of irritant atmosphere components significantly alters susceptibility to CO or vice versa. Also in real fires the conditions tending to produce high concentrations of organic irritant products need to involve relatively low temperatures, and do not normally produce high yields of CO, while under conditions promoting CO formation (high temperature flaming atmospheres) most organic irritants are destroyed [12].

We therefore believe that in real fires direct toxic interactions between the three main factors CO, HCN and irritants are unlikely and that when estimating possible incapacitative effects at particular stages of fire atmosphere development each factor should be considered separately.

Thus during the early stages of exposure to many fires the most important factor causing incapacitation and delaying escape may be the effects of irritation on the eyes and upper respiratory tract, while if escape is delayed for this or other reasons the accumulation of CO in the blood may become the main cause of incapacitation and, as it undoubtedly is in many fires, of death. HCN, when present in fire atmospheres derived from nitrogen-containing polymeric materials, may also delay escape until CO uptake eventually causes death (see below). In other situations such as long term smouldering fires where the victim is sleeping, or high temperature fires consisting of non-nitrogeneous material, CO may be the only toxic factor of importance.

It must also be recognised that if a victim encounters an oxygen depleted atmosphere then the narcotic effect of CO and/or HCN is likely to be enhanced, and that the presence of high carbon dioxide concentrations will tend to increase the rate of uptake of toxic products due to hyperventilation and even produce narcosis if present in concentrations above 10 percent.

Specific Patterns of Toxicity Relevant to Human Exposure in Fire

Having identified the three major toxic factors in atmospheres of thermal decomposition products it is relevant to consider in more detail the nature and importance of each one with regard to humans in fires, particularly the effects of escape capability.

Carbon Monoxide

Carbon monoxide must still come very high on the list of major toxic fire components. It is always present in fires, often at very high concentrations, such as percent levels (normally less than 10% but 5 to 8% may occur in severe fires), which would be fatal within a few minutes. At lower concentrations, such as the 1000-2000 ppm range used for this study the onset of CO intoxication is slow. Because of the way in which CO binds to the blood, it is usually some time before CO taken into the body builds up to a sufficiently high carboxyhaemoglobin (COHb) level to have any noticable incapacitative effect, so that clear signs of serious incapacitation were not seen until towards the end of the 30 minute exposure in the animals. However, the slow, insidious onset of CO intoxication may result in a victim being unaware of the predicament until he suddenly reaches the catastrophe stage where normal body functions can no longer be maintained, and the victim passes rapidly into a state of severe incapacitation and semiconsciousness. Whether or not such severe incapacitation occurs will depend upon the amount of CO inhaled (i.e. the COHb level) and also upon how active the victim is. Thus with the animals used in this study, which were sitting in chairs, it was noticed that when they were very quiet and relaxed they tended to reach the end of the exposure with only minor signs of incapacitation, while if they were at all active, especially towards the end of an exposure, they were likely to pass into a state of semiconsciousness. In more recent work with CO, monkeys have been placed in a chamber where they are active and free to move about, and trained to perform a behavioural task involving a certain amount of exercise and the application of psychomotor skills. In this situation the animals were found to be severely intoxicated after a 30 minute exposure to 1000 ppm CO with COHb levels approaching 30%, whereas 30 minute exposures to 2000 ppm, producing COHb levels of approximately 40% were normally necessary to produce severe intoxication in the chairbound animals. In man severe signs including collapse have been reported at COHb levels in the 30%-40%

COHb range [20] and it is likely that an active man would be seriously affected at COHb levels in the 25%-30% range, as were the monkeys. Calculations indicate that an average man engaged in light physical work in an atmosphere of 1000 ppm CO could achieve hazardous levels of 21%-37% COHb within 30 minutes, and under more extreme condition of the kind likely to be encountered in a fire, a victim could in theory achieve a level of 34% COHb within 15 minutes, even at the low atmosphere concentration of 1000 ppm CO [21,22].

Hydrogen Cyanide

With HCN the situation is rather different from that with CO, since HCN is not buffered by the blood, but causes a rapid tissue hypoxia, and also because the stimulatory effect of HCN on respiration causes the positive feedback hyperventilation described earlier. As a result of these two factors HCN is particularly dangerous because of its rapid knock-down effect, and because it is approximately 20 times more toxic than CO in terms of its ability to produce incapacitation. Low HCN levels in the 100-200 ppm range could therefore be extremely hazardous in a fire situation.

From the results with nitrogen containing polymers, HCN would be expected to be a potential factor in causing incapacitation and even death in some fires. High concentrations of HCN have been detected in large scale fire tests involving industrial loads of flexible polyurethane foam [6] and in the blood of some fire victims in the Strathclyde study [8], although high blood cyanide levels were usually associated with fatal levels of carboxyhaemoglobin so that it was difficult to determine the exact contribution of cyanide to incapacitation and death. This present study has established a useful relationship between atmosphere HCN level and time to incapacitation, with incapacitation occurring at blood cyanide levels of approximately 100 μM/litre (80-198 μM/litre). However, no relationship has been found between venous blood cyanide and atmosphere HCN concentrations over a 30 minute exposure, nor a clear relationship between venous blood cyanide and time to, or degree of, incapacitation. It may be that blood cyanide measurements are poor predictors of the degree of incapacitation caused by cyanide and that the *rate* of uptake during the hyperventilatory episode is more closely related to incapacitation and subsequent death than is the actual blood level after a period of exposure. The incapacitative effects of cyanide on the brain could be extreme during the hyperventilatory episode for a number of reasons. Apart from the direct hypoxic effect of cyanide on the cerebral tissues, the hyperventilation may cause a cerebral artery constriction resulting from a fall in the arterial partial pressure of carbon dioxide (PCO_2), which would exacerbate the hypoxia [23]. Also the circulatory failure observed in the animals at this point could further reduce cerebral blood supply. In short exposure to HCN at approximately 200 ppm could therefore result in a rapid incapacitation without producing a high blood cyanide concentration in the body, especially some hours later, so that the possibility exists that HCN may be more important in real fires than is suggested by the pathological evidence obtained so far [8].

Irritance

Generally speaking all mid-range and low temperature atmospheres (300 °C–600 °C) were rich in smoke and particulates and many were found to be irritant to some degree, especially under oxidative conditions. This is not to say that all smoke particles are necessarily irritant, for example the pyrolytic decomposition of polypropylene produced a dense but innocuous atmosphere, or that irritant products do not occur in the vapour phase of the smoke. It seems to be more that the conditions which favour the generation of smoke also favour the generation of a wide range of molecular products including for example TDI monomer or styrene monomer in specific pyrolytic atmospheres, or formaldehyde and acrolein in oxidized hydrocarbon atmospheres.

An important feature of the results from exposures to irritant atmospheres is that the responses of the primates showed considerable differences from those of the usual rodent animal models. In both animals and man, the characteristic response to airborne chemicals which are irritant to the upper respiratory tract is a decrease in respiratory rate, while the characteristic response to irritants penetrating into the lung itself is an increase in the respiratory rate, generally accompanied by a decrease in tidal volume [24]. Which of these effects predominates depends upon a number of factors, such as the physical characteristics of the aerosol, aqueous solubility of the irritant, test species and the duration of the insult. The upper respiratory tract response in rodents, a decrease in respiratory rate, has been used as a quantitative measure for evaluating the sensory irritance of airborne chemicals and has been applied to atmospheres of combustion products. In rodents the depressive effect on respiration, which tends to protect them against the effects of exposure, can last for some time (10–30 minutes) in irritant atmospheres but in primates the work has shown a mixture of upper and lower respiratory tract irritance. The initial, upper respiratory tract response, which lasted for 1–2 minutes, consisted of a decrease in respiratory rate but was soon replaced by hyperventilation consisting of an increased respiratory rate and tidal volume, which lasted for the duration of the exposure.

At high concentrations the initial decrease in respiration was very transient and the hyperventilation began within the first minute. It therefore appears that with smoke atmospheres in primates the upper respiratory tract response is rapidly adapting and gives way to a slowly adapting lung irritant response which is the dominant effect over a 30 minute exposure period. An important consequence of this response is that the hyperventilation is likely to increase the rate of uptake of irritants into the lung and lead to a more rapid incapacitation. It should also increase the rate of uptake of other toxic products such as CO or HCN, thereby hastening incapacitation from these sources.

Another feature of lung irritance is the longer term effect, which occurred following exposures to the atmosphere produced by the pyrolysis of flexible polyurethane foam at 600° C, and also by the non-flaming oxidative decomposition of polypropylene at 450° C. The response consisted of dyspnoea 12-24 hours after exposure, and in 2 cases (one from each polymer) an acute lung reaction occurred, which consisted of extensive pulmonary oedema and alveolar congestion, with an inflammatory cell exudate in the alveolar wall and alveolar space. These two atmospheres also produced the most marked signs of lung irritation during exposure, causing the greatest percentage increase in respiratory minute volume of all the atmospheres studied.

The Applications of Combustion Toxicity Research and the Development of Standard Toxicity Tests

The use of 'standard' test procedures for assessing the toxicity of decomposition and combustion products of polymeric materials (using simple laboratory equipment to generate the atmospheres with biological procedures (rodents) for assessing toxicity) have attracted considerable attention particularly in West Germany and USA, and proposed methods have been discussed [1].

Objections to the idea of tests of this kind have been put forward on the grounds that tests involving animals would be difficult to perform, and that it would be even more difficult to extrapolate from effects in rats and mice to those likely in man. Animal models are used widely in inhalation toxicity evaluation studies and within certain limitations good predictions of human acute toxicity are possible from rodent LC_{50} models [25]. The evaluation of atmospheres of combustion products should therefore be relatively easy, particularly if a simple test of time to incapacitation is included and an assessment of irritance. Indeed the results of our work would suggest that intoxicating effects can be fairly well predicted without the use of animal exposures if the CO and HCN concentrations of the atmosphere are known. The main usefulness of animal tests would then be to measure irritance, which cannot be predicted fully by chemical analysis alone and for the detection of unusual or supertoxic effects.

The major difficulty with standard combustion toxicity tests arises not from the animal model, but from the fact that the range of products given off from even a single polymeric material can very considerably depending upon the conditions under which the material is decomposed. Real fires are so complex that the variations in the overall toxicity of the atmospheres produced by even one material involved in different fires are likely to be much greater than the differences in toxicity per se between different individual materials. Our results indicate that it may be necessary to carry out a number of toxicity evaluations under different temperatures and conditions of oxidation to provide guidance about the range of possible incapacitative and toxic effects likely from a material involved in fires. It is therefore difficult to see how a simple laboratory fire model can provide an accurate reflection of actual fire conditions, and since any biological assessment of toxicity is only as good as the exposure model used to produce the atmosphere, precise measurements and ranking of the toxicity of different materials by simple procedures which are nevertheless expensive and require the use of large numbers of animals may give rise to very misleading information.

In our view the main potential for mitigating the injury and death in fires lies with overall hazard assessment. In carrying out this, albeit difficult, exercise, the typical scenario of ignition, fire growth, smoke and toxicity is not only the sequence of the fire itself, but also in many cases the sequence of mitigation. Measures which can improve the resistance to ignition, slow down the rate of burning and possibly reduce smoke obscuration are important in reducing the overall hazard in fires and currently likely to be more effective in terms of the saving of life in fires than combustion toxicity controls.

Simple combustion toxicity tests are therefore best placed in a research rather than regulatory/control category, providing knowledge for product development, and as an adjunct to other studies of large scale fire models and real fires in attempts to understand and to mitigate the processes leading to incapacitation and death in fire victims.

CONCLUDING REMARKS

The work has shown clearly the complex nature of the decomposition and combustion products of polymeric materials but the major point of great importance which has emerged is that the basis toxic effects seem to be relatively simple with the toxicity of individual atmospheres dominated by effects arising from carbon monoxide and/or hydrogen cyanide (narcotic and potentially toxic effects) or products with high irritancy.

This understanding now means that we have a much clearer picture of the likely effects of fire atmospheres, particularly at sublethal levels, with relevance to the many thousands of persons exposed each year in the UK to fires. An important inference from the work is that much more can now be done with analytical equipment (albeit sophisticated) especially in the assessment of narcotic-toxic atmospheres. Looking to the future there is also optimism in that the difficult area of irritancy may also be more amenable to analytical-chemical support than was first realised, as the knowledge gained from this present study is implemented.

Materials and composite items (particularly those of synthetic rather than natural origin) are currently causing a great deal of concern because of fears that decomposition and combustion may give rise to highly toxic products in fires. Knowledge of the simple classification of the atmospheres for toxicological purposes and the additional information which may now be possible from analytical chemistry is of relevance to all concerned with the difficult area of the mitigation of the toxic hazards in fires.

ACKNOWLEDGEMENT

This paper forms part of the work of the Fire Research Station, Building Research Establishment, Department of the Environment. It is contributed by permission of the Director, Building Research Establishment.

REFERENCES

1. W.D. Woolley and P.J. Fardell, "Basic Aspects of Combustion Toxicology," Conference on Smoke and Toxic Gases from Burning Plastics, London, 6-7 January 1982. QMC Industrial Research Ltd., London, p. 8/1-8/47. *Fire Safety Journal* Vol. 5 (1982) p. 29.
2. W.A. Harland and R.A. Anderson, "Causes of Death in Fires," Conference on Smoke and Toxic Gases from Burning Plastics, London, 6-7 January 1982. QMC Industrial Research Ltd., London. p. 15/1-15/19.
3. E.P. Radford, International Symposium on Physiological and Toxicological Aspects of Combustion Products. University of Utah. (1974).
4. S.E. Chandler and R. Baldwin, "Furniture and Furnishings in the Home, Some Fire Statistics," *Fire Mater.* Vol. 1 (1976) p. 76.
5. U.K. Fire Statistics, London, Home Office, published annually (1979).
6. W.D. Woolley and P.J. Fardell, "The Prediction of Combustion Products," *Fire Research* Vol. 1 (1977) p. 11.
7. G. Kimmerle, "Aspects and Methodology for the Evaluation of Toxicological Parameters During Fire Exposure," *JFF/Combustion Toxicology* Vol. 1 (1974) p. 4.
8. W.A. Harland and W.D. Woolley, "Fire Fatality Study," University of Glasgow. Building Research Establishment Information Paper IP 18/79 (1979).
9. J.H. Petajan, K.J. Voorhees, S.C. Packham, R.C. Baldwin, I.N. Einhorn, M.L. Grunnet, B.G. Dinger and M.M. Birky, "Extreme Toxicity from Combustion Products of a Fire-Retarded Polyurethane Foam," *Science*, Vol. 187 p. 742 (1975).
10. H.J. Klimisch, "Concept and Experience with the Standardisation of Test Methods According to DIN 54 436 for Determining the Relative Acute Inhalation Toxicity of Thermal Decomposition Products," Conference on Smoke and Toxic Gases from Burning Polymers, London 6-7 January 1982. p. 9/1-9/30.

11. D.A. Purser, P. Grimshaw and K.R. Berrill, "The Role of Hydrogen Cyanide in the Acute Toxicity of the Pyrolysis Products of Polyacrylonitrile," In Press.
12. W.D. Woolley, S.A. Ames and P.J. Fardell, "Chemical Aspects of Combustion Toxicology of Fires," *Fire Mater.* Vol. 3 (1979) p. 110.
13. W.D. Woolley, "Nitrogen-Containing Products from the Thermal Decomposition of Flexible Polyurethane Foam," *Br. Polym. J.* Vol. 4, p. 27.
14. J.F. Nunn, *Applied Respiratory Physiology*, Butterworth, 1969. p. 156.
15. W.D. Woolley, P.J. Fardell and I.G. Buckland, "The Thermal Decomposition Products of Rigid Polyurethane Foams under Laboratory Conditions," Fire Research Station, Fire Research Note No. 1039.
16. D.A. Purser and P. Grimshaw, "The Incapacitative Effects of Exposure to the Thermal Decomposition Products of Polyurethane Foams," Interflam '82 Conference, Guildford, 30 March-1 April 1982. *Fire Mater.* In Press.
17. R.D. Lynch, on the Non-Existence of Synergism Between Inhaled Hydrogen Cyanide and Carbon Monoxide. Fire res. Station, Fire Res. Note 1035. Borehamwood (1975).
18. B.R. Pitt, E.P. Radford, G.H. Gurtner and R.J. Traystman, "Interaction of Carbon Monoxide and Cyanide on the Cerebral Circulation and Metabolism," *Arch. Environ. Hlth.* Vol. 34 (5) (1979) p. 354.
19. E.A. Higgins, V. Fiorca, A.A. Thomas and H.V. Davis, "Acute Toxicity of Brief Exposures to HF, HC_1, NO_2 and HCN With and Without CO," *Fire Technology*, Vol. 8 (1972) p. 120.
20. R.D. Stewart, J.E. Peterson, E.D. Baretta, H.C. Dodd and A.A. Herrmann, "Experimental Human Exposure to Carbon Monoxide," *Arch. Environ. Hlth.* Vol. 21 (1970) p. 154.
21. R.D. Stewart, "The Effects of Carbon Monoxide on Man," *JFF/Combustion Toxicology* Vol. 1 (1974) p. 167.
22. J.E. Peterson and R.D. Stewart, "Predicting the Carboxyhaemoglobin Levels Resulting from Carbon Monoxide Exposures," *J. app. Physiol.* Vol. 39 (1975) p. 633.
23. J.B. Brierley, A.W. Brown and J. Calverley, "Cyanide Intoxication in the Rat: Physiological and Neuropathological Aspects," *J. Neurol. Neurosurg. Psychiat.* Vol. 39 (1976) p. 129.
24. D.G. Clark, S. Buch, J.E. Doe, H. Frith and D.H. Pullinger, "Bronchopulmonary Function: Report on the Main Working Party," *Pharmac. Ther.* Vol. 5 (1979) p. 149.
25. N. I. Sax, *Dangerous Properties of Industrial Materials* 4th Edition, Van Norstrand Reinhold Co., 1975.

5

Influence of Heating Rates on the Toxicity of Evolved Combustion Products: Results and a System for Research

SHAYNE COX GAD and ANN C. SMITH

Allied Corporation[1]
Department of Toxicology
P.O. Box 1021R
Morristown, NJ 07960

ABSTRACT

The significance of the rate of heating of materials to the nature and toxicologic consequences of combustion products formed by natural and man-made products was evaluated using a system designed to allow exact control and reproducibility of this variable. Using this system, the decomposition products of Douglas Fir, Hem Fir, and a high density polyethylene were characterized in terms of gases evolved, lethality, and ability to incapacitate at sample heating rates of 20, 30, 40, and 50°C per minute. For all three materials, the rate of heating was found to have marked influences on both the decomposition products and their toxicologic impact. This influence was not such as to be explainable as just a decomposition rate phenomenon.

INTRODUCTION

AS SHOULD BE apparent to those involved in the field by this point, combustion toxicology has grown and evolved to the point that it is approaching a transformation. To this point, as is the case with much of toxicology, the field has progressed in the manner of a descriptive science—data have been collected which act to describe what happens under a variety of situations, but not to explain why things happen. This second step—the development of models which explain why things

---

[1]Authors current address but not site of performance of bulk of experimental works.

happen and which therefore serve to predict as yet unobserved happenings—is the mechanistic phase of a science. Combustion toxicology is on the verge of its mechanistic phase. In order to move the field to the mechanistic phase, there are several questions to be answered.

The first question which must be addressed is that of the relationship between the large scale (or "real") fire and the various laboratory models which are currently being employed. Measurements of what goes on in the different regions of a real fire and studies of how particular materials behave in a full scale situation have been made [1,2,3]. But only very few attempts have been made to study the biological effects of large scale fire atmospheres, and these have largely not been published. Some of the laboratory scale models attempt to address this indirectly by incorporating evaluations of several diverse thermolysis conditions (flaming, nonflaming and 440 °C for example) into the evaluations. But a true validation of any of the proposed test systems will require a series of parallel experiments (large scale and laboratory scale) which directly compare results of defined loads of test materials. Only this will provide a verification of where the results of the laboratory tests are valid predictors and where they are not. The mathematical modeling efforts for fires which have been a primary objective of much of the large scale fire testing to date have (unfortunately) progressed only to the stage of accounting for very simple fire situations. These modeled situations are even less real life than the laboratory scale tests themselves.

The second set of basic questions which must be answered before the field can be advanced to the mechanistic phase are the relationships between toxicity and variations in (A) restrictions in available oxygen during decomposition and (B) alterations in heating rates. Some work has been done on both of these questions but neither has been definitively addressed.

Originally, it was believed that the most dangerous thermolysis state in terms of the production of toxic products was the smoldering (or limited air available) condition. Very little work has directly addressed this point. Hilado and Brauer [4] did present some evaluations of the effects of variations in air flow through their system, which does not quite address the problem. Other work [5,6] supports the view that limited oxygen in a thermolysis environment does lead to increased production of CO as opposed to CO_2. But a complete addressal is required.

The question of the influence of heating rates has been more widely recognized and addressed. Cornish et al. [7] evaluated the differences in toxicity of products of decomposition when the samples were slowly heated at 5 °C/minute as opposed to a rapid heating and found major differences in toxicity. Hilado et al. [8] likewise studied LC_{50} values of 12 materials as they were affected by alterations in the final sample temperature and in heating rate. They believed that simple thermo-dynamics (rate phenomena) explained the wide variations that were observed. Hoff *et al* [9] have additionally recently published some work on these aspects with results which do not support a purely kinetic relationship.

A wide variety of testing models have been developed and described in the literature [2]. None of these, however, allows a direct addressal of the questions raised above. The CMIR (Carnegie Mellon Institute of Research) system [10] and the protocol described here do.

Using the CMIR system, some years ago a series of studies (not previously reported in the literature) was conducted on the "ramping" effect—that is, the alterations in results produced by different heating rates. The equipment utilized was able to reproducibly control sample heating rates over a very wide range.

There is another major area where data, though of obvious significance, are not being collected with current research designs. This is the impact of smokes and aerosols—what levels of what types are being generated, and what are their biological effects? Further pursuit of these questions is beyond the scope of the current article.

MATERIALS AND METHODS

From 1977 to 1979 a system and protocol for evaluating combustion toxicity of polymers was developed at what was then the Chemical Hygiene Fellowship of CMIR.[2] This development was conducted by T. F. Brecht, J. B. Reid, and S. C. Gad.

The furnace utilized is shown in Figure 1 and diagrammed in Figure 2. A focused infra red heater serves to heat a sample in a quartz plate on a thermogravimetric analysis (TGA) device. A thermocouple located in the sample serves to monitor sample temperature and can control in the 150 °C to 1500 °C (± 5 °C) range. Alternatively, the system can control by means of heat flux from the IR heater. The furnace, as shown in Figure 3, is not directly in the animal exposure chamber. Rather, it is connected by a series of insulated glass tubes which serve to make the entire system closed. An in-line sealed (only the blades are open to the test atmosphere) fan acts to circulate and mix the chamber contents, and warming tapes around the glass tubing prevent condensation and collection of aerosols. Visual inspection of system lines after tests and the performance of mass balance calculations have verified that only minimal aerosol deposition occurs—usually below the limits of detection.

A Data Trak 5600 microprocessor controls the entire system, allowing complete and precise reproducibility and the ability to alter sample

---

[2]Now known as Bushy Run Research Center, Export, PA.

Figure 1. Microprocessor controlled furnace for CMIR thermolysis toxicity testing system.

temperature in the desired manner (5 °C/min, 10 °C/min, etc). The furnace is also configured to allow complete control over the initial atmosphere of the furnace (by adding O_2, N_2, or CO_2), and a propane torch acts as an ignition source when necessary. Concentrations are calculated based on the weight of sample volatilized.

Analytical monitoring in this system is based primarily on continuous measurement and recording of O_2, CO, CO_2, NO/NO_X (by IR spectrophotometry[3]), total hydrocarbons (by flame photometry[3]), and HCN (by a specific electrode[4]). Periodic measurements of other, special interest gases can be made by draw tubes.

The animal exposure system is a glass cylinder with stainless steel ends, having a motor driven activity wheel suspended on a stainless steel shaft through the chamber. The volume of the complete system is 55 liters.

[3]Beckman Instruments; O-Model F3M3; CO-Model 865; CO_2-Model 864; NO/NO_X-Model 402; HC-Model 402.
[4]Dow.

Six male Sprague Dawley rats were utilized for each 30 minute exposure. Incapacitation was considered to occur once an animal was incapable of walking on the wheel, but rather tumbled. A high intensity lamp placed behind the chamber provided the opportunity to observe animals when even a heavy smoke was generated. A disadvantage was that this incapacitation model can be called a subjective one, and indeed it requires a degree of observer training. One approach to adding an objective measure to the system was monitoring the torque or drag on the turning wheel, but this effort has not been pursued to completion. Endpoints from the animal's system were: time to incapacitation, lethality and (with blood taken from one animal, by heart puncture after the exposure), COHb, O₂Hb, and total Hb.

The protocol employed for the current study entailed putting a weighed sample of test material into the quartz ignition plate for each trial. Six male Sprague Dawley rats (120-180 g, from Hilltop Research Animals, Scottsdale, PA) were loaded into the exposure system, and air circulation within the closed system was initiated. Starting from a baseline temperature (25 ± 2 °C), sample temperatures were raised at a set rate of either 20, 30, 40, or 50 °C per minute for 30 minutes. Time to

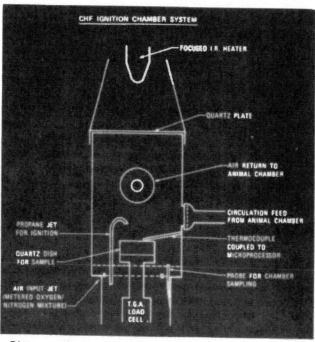

Figure 2. Diagram of microprocessor controlled furnace and decomposition chamber with feed back control sensors.

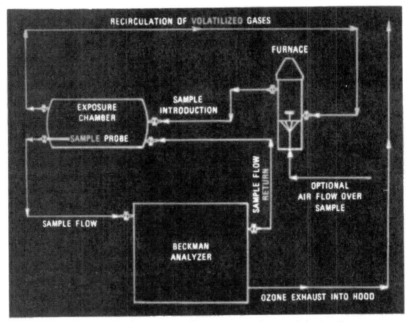

Figure 3. Schematic diagram of CMIR combustion toxicity testing system.

incapacitation of animals, sample temperatures, concentrations of gases (O_2, CO_2, CO, total hydrocarbons, NO_X, and HCN), and animal chamber temperatures were continuously monitored and recorded. In accordance with the 1977 National Academy of Science guidelines [11], oxygen levels for animals never went below 18%, and animal chamber temperatures never exceeded 30 °C. Measureable levels of nitrogen oxides (0.5–5.3 ppm) were seen, but only with the woods in the 50 °C/minute runs. Where deaths and/or incapacitations occurred, LC_{50} and IC_{50} values were calculated.

Nominal toxicant levels were determined by measuring the amount of test material sample weight volatilized, then dividing this by the volume of the system.

LC_{50} and IC_{50s} were calculated by the probit method [12], while other methods were performed as described previously [13].

RESULTS

Data from the three sets of studies are presented in Tables 1 (Douglas Fir), Table 2 (Hem Fir), and Table 3 (High Density Polyethylene).

All the data cells are not complete due to some instrument difficulties. The most common of these (solved early in the study) was a

Table 1. Douglas Fir heating rate studies.

Sample Heating Rate [1]	Fuel Load [2]	Nominal Toxicant Concentration [3]	Mortality Results	Incapacitation Results	Time To First Incapacitation	Maximum CO_2,%	Maximum CO(%)	Maximum HC(PPM)
20°	2 gm	18.6 mg/L	0/6	0/6	N/A	3.1	0.25	240
	3 gm	28.1 mg/L	0/6	0/6	N/A	*	*	*
	3 gm	41.3 mg/L	0/6	6/6	24'-38"	1.8	0.60	60
	4 gm	44.4 mg/L	4/6	6/6	27'-18"	3.1	0	580
		LC 50 = 43.0 mg/L						
		IC 50 = 33.5 mg/L						
30°	1 gm	10.5 mg/L	0/6	0/6	N/A	*	*	*
	2 gm	19.4 mg/L	0/6	3/6	26'-54"	3.6	0.25	660
	3 gm	33.0 mg/L	2/6	6/6	*	*	*	*
	4 gm	44.0 mg/L	4/6	6/6	24'-38"	2.3	0.55	43
		LC 50 = 38.5 mg/L						
		IC 50 = 19.9 mg/L						
40°	2 gm	12.8 mg/L	0/6	0/6	N/A	3.5	0.25	*
	2 gm	13.6 mg/L	0/6	0/6	N/A	2.9	0.10	*
	2 gm	17.7 mg/L	0/6	0/6	N/A	2.8	0.15	*
	3 gm	32.8 mg/L	0/6	6/6	21'-29"	2.9	0.60	*
	4 gm	55.7	5/6	6/6	21'-03"	1.8	0.65	*
		LC 50 = 39.3 mg/L						
		IC 50 = 19.6 mg/L						

(continued)

71

Table 1. (continued)

Sample Heating Rate[1]	Fuel Load[2]	Nominal Toxicant Concentration[3]	Mortality Results	Incapacitation Results	Time To First Incapacitation	Maximum CO_2%	Maximum CO(%)	Maximum HC(PPM)
50°	1 gm	7.0 mg/L	0/6	0/6	N/A	2.8	0.10	105
	2 gm	18.6 mg/L	0/6	2/6	28'-54"	4.2	0.12	255
	2 gm	25.6 mg/L	6/6	6/6	20'-20"	3.4	0.15	330
	3 gm	31.2 mg/L	6/6	6/6	20'-15"	3.9	0.9	375
	4 gm	50.4 mg/L	6/6	6/6	19'-37"	3.7	1.40	510
	6 gm	85.4 mg/L	6/6	6/6	19'-40"	3.6	1.80	590
				LC 50 = 21.5 mg/L				
				IC 50 = 14.8 mg/L				

Notes: [1] °C per minute sample temperature was increased from 25°C to autoignition.
[2] Weight of sample loaded into system.
[3] Total weight loss of sample divided by system volume.
[4] All gas levels are peak values measured — not integrated totals.
* = Values not measured.
HC = Hydrocarbons.

72

Table 2. Hem Fir heating rate studies.

Sample Heating Rate[1]	Fuel Load[2]	Nominal Toxicant Concentration[3]	Mortality Results	Incapacitation Results	Time To First Incapacitation	Maximum CO₂,%	Maximum CO(%)	Maximum HC(PPM)
20°	2 gm	20.4 mg/L	0/6	0/6	N/A	3.3	0.25	230
	3 gm	31.2 mg/L	0/6	4/6	28'-36"	2.9	0.42	290
	4 gm	49.3 mg/L	6/6	6/6	23'-27"	2.7	0.58	460
	5 gm	60.7 mg/L	6/6	6/6	19'-24"	1.9	0.66	715
				LC 50 = 45.4 mg/L				
				IC 50 = 30.0 mg/L				
30°	1 gm	9.8 mg/L	0/6	0/6	N/A	2.9	0.12	130
	2 gm	18.4 mg/L	0/6	3/6	26'-58"	3.7	0.21	260
	3 gm	36.7 mg/L	2/6	6/6	25'-17"	3.2	0.47	180
	4 gm	47.7 mg/L	6/6	6/6	23'-47"	2.6	0.64	390
	5 gm	59.8 mg/L	6/6	6/6	22'-18"	2.1	0.83	540
				LC 50 = 33.2 mg/L				
				IC 50 = 18.6 mg/L				
40°	1 gm	10.2 mg/L	0/6	0/6	N/A	3.3	0.21	150
	2 gm	19.4 mg/L	0/6	2/6	26'-53"	2.8	0.24	250
	3 gm	37.4 mg/L	4/6	6/6	21'-13"	2.4	0.46	410
	4 gm	52.2 mg/L	6/6	6/6	21'-02"	1.7	0.85	530
	5 gm	65.9	6/6	6/6	20'-13"	1.5	1.12	630
				LC 50 = 32.8 mg/L				
				IC 50 = 18.7 mg/L				

(continued)

73

Table 2. (continued)

Sample Heating Rate[1]	Fuel Load[2]	Nominal Toxicant Concentration[3]	Mortality Results	Incapacitation Results	Time To First Incapacitation	Maximum CO_2%	Maximum CO(%)	Maximum HC(PPM)
50°	1 gm	10.4 mg/L	0/6	0/6	N/A	2.9	0.13	210
	2 gm	21.6 mg/L	2/6	6/6	27'-49"	3.9	0.32	330
	3 gm	38.5 mg/L	6/6	6/6	20'-19"	2.7	0.74	490
	4 gm	53.7 mg/L	6/6	6/6	19'-43"	1.7	1.24	*
	5 gm	67.8 mg/L	6/6	6/6	19'-30"	1.6	1.70	*
				LC 50 = 21.5 mg/L				
				IC 50 = 14.8 mg/L				

Notes: (1) °C per minute sample temperature was increased from 25°C to autoignition.
     (2) Weight of sample loaded into system.
     (3) Total weight loss of sample divided by system volume.
     (4) All gas levels are peak values measured – not integrated totals.
     * = Values not measured.
     HC = Hydrocarbons.

74

Table 3. High density polyethylene heating rate studies.

Sample Heating Rate[1]	Fuel Load[2]	Nominal Toxicant Concentration[3]	Mortality Results	Incapacitation Results	Time To First Incapacitation	Maximum CO_2%	Maximum CO(%)	Maximum HC(PPM)
20°	2 gm	9.8 mg/L	0/6	1/6	28'-15"	2.7	0.08	120
	3 gm	14.6 mg/L	0/6	6/6	27'-48"	3.0	0.10	175
	4 gm	19.9 mg/L	0/6	6/6	27'-17"	3.2	0.12	252
	5 gm	24.8 mg/L	2/6	6/6	26'-18"	3.5	0.56	360
	6 gm	29.7 mg/L	6/6	6/6	26'-19"	4.3	0.65	390
				LC 50 = 25.2 mg/L				
				IC 50 = 10.6 mg/L				
30°	1 gm	8.9 mg/L	0/6	2/6	28'-12"	2.9	0.12	130
	2 gm	19.7 mg/L	0/6	6/6	26'-34"	3.1	0.24	242
	3 gm	28.2 mg/L	6/6	6/6	24'-18"	3.4	0.60	375
	4 gm	32.4 mg/L	6/6	6/6	24'-13"	3.2	0.89	480
	5 gm	39.6 mg/L	6/6	6/6	23'-37"	1.6	1.45	560
				LC 50 = 24.7 mg/L				
				IC 50 = 9.6 mg/L				
40°***	0.5 gm	12.4 mg/L	0/6	6/6	27'-02"	2.9	0.08	255
	1 gm	25.3 mg/L	0/6	6/6	24'-17"	3.3	0.10	520
	2 gm	50.8 mg/L	2/6	6/6	22'-19"	3.6	0.23	•

(continued)

Table 3. (continued)

Sample Heating Rate[1]	Fuel Load[2]	Nominal Toxicant Concentration[3]	Mortality Results	Incapacitation Results	Time To First Incapacitation	Maximum CO_2%	Maximum CO(%)	Maximum HC(PPM)
	3 gm	83.4 mg/L	6/6	6/6	20'-12"	3.9	0.69	*
	4 gm	90.5	6/6	6/6	18'-04"	4.5	1.02	*
				LC 50 = 48.4 mg/L				
				IC 50 = <12.4 mg/L				
50**°	0.5 gm	19.6 mg/L	0/6	6/6	24'-12"	3.2	0.12	320
	1 gm	39.3 mg/L	0/6	6/6	22'-19"	3.7	0.26	*
	2 gm	64.5 mg/L	6/6	6/6	17'-18"	4.5	0.73	*
	3 gm	91.2 mg/L	6/6	6/6	16'-15"	4.6	1.20	*
	4 gm	94.3 mg/L	6/6	6/6	15'-29"	4.9	1.40	*
				LC 50 = 51.2 mg/L				
				IC 50 = <19.6 mg/L				

Notes: [1]°C per minute sample temperature was increased from 25°C to autoignition.

[2]Weight of sample loaded into system.

[3]Total weight loss of sample divided by system volume.

[4]All gas levels are peak values measured — not integrated totals.

* = Values not measured.

HC = Hydrocarbons.

** = Flaming of samples occurred early in temperature sequence.

76

clogging of the filter system on the hydrocarbon analyzer, leading to the loss of a number of determinations on this measure.

Several patterns are clear in these data. First, as the rate of sample heating increases, the LC_{50} and IC_{50} both decrease for woods but increase for HDPE. Oxygen levels and animal temperatures did not vary in any manner correlated with sample heating rate or to influence the observed outcomes.

Second, the nominal toxicant level (that is, the amount of sample consumed) did not increase linearly with sample heating rate across the whole range of fuel loads and materials. Rather, at the higher fuel loads (particularly at lower heating rates), there appears to be a limitation to how much sample will be consumed (most markedly in the HDPE). It was observed in these cases that an ash or crust layer tends to develop on the top of a sample and this crust appears to impede further decomposition of samples.

Third, it is also clear that carbon dioxide, carbon monoxide and total hydrocarbons produced are not linear with fuel load, and that heating rates affect the proportions of these produced. The amount of CO_2 produced actually declines with larger sample loads (except where flaming occurred), while both CO and total hydrocarbons increase (though not linearly) as sample load increases. As sample heating rate increases, the proportion of CO produced per fuel load increases.

Fourth, flaming (as occurred in the higher heating rates with high density polyethylene) produces an entirely different set of conditions than does nonflaming. LC_{50s} and IC_{50s} actually decline with heating rates here, and the proportion of carbon monoxide to fuel load is lower than in the nonflaming tests. Additionally, carbon dioxide production increases (though not linearly) with toxicant concentration. Oxygen levels did decline slightly with increased toxicant concentrations in the flaming trials, but remained above 18% in all cases.

Finally, it is striking in the HDPE data that incapacitation is occurring in many trials where there is no mortality. The levels of carbon monoxide present were not sufficient to account for the observed incapacitations.

CONCLUSIONS

First, under the whole range of conditions of test, the combustion toxicity of HDPE is no greater than that of wood. Under flaming conditions, incapacitation by the combustion products of HDPE can occur without death. The differential manner in which incapacitation and death occur with HDPE indicates that these two endpoints are not due to the same mechanism or agents.

Second, theremal decomposition of large fuel loads without flaming favors carbon monoxide (as opposed to carbon dioxide) production. In woods, however, there is seemingly no significant difference in potential to kill or incapacitate between the results of flaming and nonflaming decomposition.

Lastly, heating rates appear to have profound effects on the toxicologic potential of thermal degradation products. This influence is not limited to a matter of rate of decomposition and should be studied further.

Although the results of this study raise more questions than they answer, the above points seem clear.

The apparent underlying principles, however, are more important. The conditions under which a sample of material is decomposed must be well defined and rigidly controlled, for the presence or absence of a flame leaves too much variability. Furthermore, the relation of the surface area of a material to its total mass or volume (portrayed here by the weight of fuel load) noticeably alters, by varying the availability of oxygen for decomposition reactions, the nature of the products produced and thus the hazard represented by the material in a fire. Finally, the full range of conditions in a fire should be viewed as representing an opportunity for a full spectrum of thermal decomposition products to develop. This requires that any laboratory scale system which will properly assess the potential hazard of materials in fire must evaluate a range of thermal decomposition conditions.

REFERENCES

1. Grand, A.F., Kaplan, H.L., and Lee, G.H., "Investigation of Combustion Atmospheres in Real Building Fires," Southwest Research Institute Project Report No. 01-6067, San Antonio, Texas (1981).
2. Kaplan, Harold L., Grand, A.F., and Hartzel, G.E., *Combustion Toxicology; Principles and Test Methods*, Technomic Publishing Co., Lancaster, PA (1983).
3. Woolley, W.D., Ames, S.A., and Fardell, P.J., "Chemical Aspects of Combustion Toxicology in Fires," *Fire Mater 3*, pp. 110-20 (1979).
4. Hilado, Carlos J., and Brauer, D.P., "How Test Conditions and Criteria Affect Pyrolysis—Gas Toxicity Findings," *Mod Plast 56*, pp. 62-64 (1979).
5. Herrington, R.M., and Story, B.A., "The Release Rate of Heat, Smoke and Primary Toxicants from Burning Materials," *J Fire Flamm 9*, pp. 284-307 (1978).
6. Woolley, W.D., and Fardell, P.J., "Basic Aspects of Combustion Toxicology." *Fire Safety Journal 5*, pp. 29-48 (1982).
7. Cornish, Herbert H., Hahn, K.J., and Barth, M.L., "Experimental Toxicology of Pyrolysis and Combustion Hazards," *Environm Hlth Perspect 11*, pp. 191-196 (1975).
8. Hilado, Carlos J., Solis, A.N., Marcussen, W.H., and Furst, A., "Effect of Temperature and Heating Rates on Apparent Lethal Concentrations of Pyrolysis Products," *J Combust Toxicol 3*, pp. 381-392 (1976).
9. Hoff, A., Jacobson, S., Pfaeffli, P., and Zitting, A., *Scand J. Work Envir Hlth 8*, Issue Supp 2 (1982).

10. Gad, Shayne C., "Toxicology of Combustion Products of Polymers, in *Flame Retardancy of Polymeric Materials, Vol. 6*, Edited by William C. Kuryla and Anthony J. Papa, Marcel Dekker, Inc., New York (In Press).
11. Committee on Fire Toxicology of the National Research Council, *Fire Toxicology: Methods for Evaluation of Toxicity of Pyrolysis and Combustion Products Report No. 2*, National Academy of Sciences, Washington, DC (1977).
12. Finney, Donald J., *Probit Analysis*, 3rd Ed. Cambridge University Press, Cambridge, England (1977).
13. Gad, Shayne C., and Weill, C.S., "Statistics for Toxicologists," *Principles and Methods in Toxicology* (A.W. Hayes, Ed.), pp. 273-320, Raven Press, New York (1982).

6

A Comparative Review of the Combustion Toxicity of Polyvinyl Chloride

ROBERT K. HINDERER

The BF Goodrich Company
Environmental Health Department
500 South Main Street
Akron, Ohio 44318

INTRODUCTION

OVER THE PAST fifteen years considerable attention has been given to the combustion toxicity of materials. This concern has largely been the result of a recognition that most fire deaths result from the inhalation of toxic gases, particularly carbon monoxide (Harland and Wooley, 1979; Halpin and Berl, 1978; and Birky et al, 1979). Although the relative importance of toxicity in the overall fire hazard of a material is not known, it is recognized to be one of the many factors, (i.e. quantity, configuration, compartment volume and surface area, ventilation, flammability and combustion properties, presence and type of ignition source, fire protection systems, building occupancy, and code enforcement) which influence fire hazard (likelihood or probability of injury).

One of the materials which has drawn particular attention over the years is polyvinyl chloride (PVC). Many early attempts to compare the combustion toxicity of PVC and other materials with natural products such as wood involved the analysis of combustion gases and their respective toxicities. While qualitative and quantitative differences were found, many materials released similar types, namely commonly known asphyxiants such as carbon monoxide (CO) and carbon dioxide (CO₂), irritants, and complex organic chemicals. In the case of PVC and wood, both materials produced CO and CO_2 and complex organics. However the subject of real contention was the production of irritants—hydrogen chloride (HCl) from PVC and acrolein, formaldehyde,

80

butyraldehyde and acetaldehyde from wood. Some contended that the higher potential quantitative yield of HCl from PVC would make it unusually toxic or many times more toxic than wood. Because acrolein and formaldehyde were, respectively, 50 and 5 times more toxic than HCl (ACGIH, 1982; Kimmerle, 1974; Darmer et al, 1974; Yuill, 1974; Henderson and Haggard, 1943; and Fassett, 1963) and because wood produced more CO than PVC, it was argued that the combustion toxicity of PVC was not likely to differ greatly from that of wood. In general, the analytical approach led to unsubstantiated allegations and was unable to predict, apriori, possible toxic interactions.

After much speculation and debate it became clear that a more objective way of evaluating and comparing the combustion toxicity of materials was needed. This recognition shifted the emphasis from extrapolation based on analytical data and pure gas toxicology to the development of a small scale bioassay procedure.

Because there have been so many claims that PVC has a "unique toxic syndrome" and that PVC is extremely or unusually toxic, comparative studies of PVC and natural products have been reviewed to determine whether the existing data supports these claims. Efforts also have been made to discuss the relevance of various methods to real world fire conditions and to discuss the possible reasons for different toxic responses in various assays.

Small Scale Combustion Toxicity Tests with PVC

Although a considerable amount of combustion research has been conducted in the last 20 years, only a relatively few studies have compared the combustion toxicity of various materials. Furthermore, most of these comparative studies have been conducted within the last six years. Because of protocol differences, the data in one study generally cannot be directly compared to that from another. Still, these studies have been extremely valuable, not only because they allow the comparison of materials under identical conditions, but because they cover a wide range of conditions.

One of the earliest studies which compared the combustion toxicity of PVC and natural products was reported by Hofmann and Oettel (1969). Similar data was also reported by Hofmann and Sand (1974). Here the authors used the German Standards Institution test, DIN 53 436, which is generally referred to as the DIN method (Kaplan et al, 1983). Combustion toxicity was evaluated by exposing rats for 30 minutes to the decomposition products from materials heated at 200, 300, 400, 500 and 600 °C. As can be seen in Table 1, the only material which produced mortality at 200 °C was celluloid, whereas at 300 °C all materials except expanded cork resulted in deaths. From this data PVC appeared to be less toxic than celluloid and leather, about as toxic

Table 1. Comparison of the combustion toxicity of PVC and natural product using the DIN method.[a]

Material[b]	% Mortality				
	200°C	300°C	400°C	500°C	600°C
PVC	0	83	92	100	100
Celluloid	100	100	100	100	100
Fir	0	72	100	100	100
Expanded Cork	N.D.	0	28	100	100
Wool	N.D.	17	100	100	100
Pinewood	N.D.	25	100	100	100
Felt	N.D.	50	100	100	100
Leather	N.D.	100	100	94	92

[a]Hofmann and Oettel, 1969.
[b]Sample Weight = 5g.

as fir and felt, and slightly more toxic than wool and pinewood[1]. Expanded cork appeared to be the least toxic.

While some of these early studies evaluated the toxicity or toxic potency of materials by varying the amount of sample, others such as Kishitani (1971) studied toxic potency by varying the exposure time. The studies reported the data as "time-to-effect," however in reality they were actually comparing responses to differences in total dose. The effect of exposure time is quite evident if one compares dimethylamine (DMA) and carbon monoxide (CO). At 4500-5700 ppm, both chemicals will produce 50% mortality in rats (Steinhagen et al, 1982 and Packham and Hartzell, 1981). The difference is that DMA requires six hours exposure to achieve this response while CO requires only 0.5 hours. Therefore, DMA is a less potent toxicant because it requires a total dose many times greater than that for CO.

In these studies by Kishitani (1971), mice were preconditioned to a constant temperature of 60°C for 48 hours and then were exposed in a 56 1 chamber to the combustion gases from various materials. Each sample was heated gradually to 740°C in 15 minutes (approximately 50°C/min). Under these dynamic conditions (2 1/min.) Kishitani observed a more rapid occurrence of death in some animals exposed to the decomposition products from cedar, fire retarded plywood, and treated finishing board. However, on the whole, the exposure duration required to cause death for these materials was not appreciably dif-

---

[1]Because of inter-and intra-laboratory test variability and because of the unknown relevance of small scale combustion toxicity test to real world toxic potential, numerical comparisons are only acceptable within the narrow confines of research and development for identifying trends. Furthermore, numerical values should be viewed only as approximations, not absolute numbers.

Table 2. Comparison of the time required for PVC and natural products to cause death in mice.[a]

Material[b]	Time to Death (min)[c]
PVC	15.0
Cedar	10.7–15.0
Plywood, fire retarded	11.0–14.5
Finishing board, melamine & Phenol impregnated paper	13.5–14.3

[a]Kishitani, 1971.
[b]Equal amounts tested and nearly equal weight loss.
[c]Expressed as the range of times observed for 5 mice.

ferent from that required for the combustion products generated with PVC (Table 2).

Subsequent studies by Kishitani and Yusa (1978) compared the amount of sample required to cause collapse (i.e. incapacitation). Using a similar, but somewhat modified combustion exposure system, these researchers exposed groups of mice to samples combusted at two fixed temperatures: 550 °C or 850 °C. Revolving cages facilitated the determination of collapse in the animals. At 850 °C the amount of sample required to cause animals to collapse was similar for Japanese cedar, luan, hardboard, untreated plywood, flame retarded plywood, and PVC foam (Table 3). The significantly higher value for PVC flexible sheet indicated that it was the least toxic of the materials tested, while

Table 3. Comparison of the amount of material required to produce collapse in mice in 2.5 to 6.0 minutes when combusted at selected temperatures.[a]

Materials	Amount of Sample Required for Collapse (grams)	
	550 °C	850 °C
Japanese Cedar	5.50–8.32[b]	3.72–5.83[b]
Luan	6.13–9.11[b]	3.47–5 21[b]
Hardboard	—	3.09–4.46[b]
Untreated Plywood	4.35–6 98[c]	1.31–2.19[b]
Plywood, Flame Retarded	3.42–4.91[b]	2.20–3.55[b]
Wool	—	0.41–0.60[b]
PVC Foam	3.45–5.45[b]	2.61–4.02[b]
PVC Flexible Sheet	—	34.5–48.60[d]

[a]Kishitani and Yusa, 1979.
[b]Collapse time = 2.5–3.5 minutes.
[c]Collapse time = 3.5–4.5 minutes
[d]Collapse time = 4.5–6.0 minutes.

wool was the most toxic. Of the materials tested at 550 °C (Japanese cedar, luan, untreated plywood, flame retarded plywood, and PVC foam), no significant difference in toxicity was noted.

Another study which compared the combustion toxicity of PVC and natural products was conducted by Cornish et al (1975). Rats were exposed in a 1500 l chamber to the combustion products of different samples. Both static and dynamic conditions were evaluated. Under static conditions, materials were decomposed at 700 °C which was reached in 1-2 minutes, and the animals were exposed for 4 hours and observed for 7 days. In the dynamic system, samples were degraded by increasing temperatures at 5 °C/min to a maximum of 700 °C and the animals were exposed for 140 minutes.

Table 4 compares the combustion toxicity of PVC, wool, red oak, and cotton under the static and dynamic conditions reported by Cornish et al (1975). In the static mode, comparisons of these materials on the basis of largest nonlethal amount (LNA), estimated LC_{50} and the 0 to 100% mortality range show good correlation. In all three of these comparisons, PVC and wool were less toxic than red oak and cotton. However, under dynamic conditions the responses were quite variable. When compared on the basis of the LNA, red oak and cotton were less toxic than wool and PVC—just the reverse of what was observed in the static system. This response was also different than what was observed by comparing the estimated LC_{50} values for each sample. While the toxicity, as measured by the LC_{50}, did not change the numerical ranking of materials in the dynamic mode, the data indicated that the toxicity of PVC, red oak, and cotton were similar. Only wool stood out as being more toxic. A comparison of the 0 to 100% mortality range also suggested that PVC, red oak, and cotton were less toxic than wool.

Another study by Boudene et al (1977) compared the combustion toxicity of PVC and Douglas fir at 400 °C and 800 °C with and without the use of water traps.[2] The toxicities for these materials varied depending on the conditions (Table 5). Without the use of water traps, PVC was comparable to, or slightly more toxic, than Douglas fir. However, when water traps were used to investigate the possibility that water (i.e. from sprinkler systems) might provide protection from toxic gases, PVC was less toxic than Douglas fir. This difference appeared to be the result of the high solubility of HCl in the water, the lower levels of CO produced by PVC than wood, and the inability of water to remove CO from the air.

While many groups have studied the combustion toxicity of materials, perhaps the most prolific work is that of Hilado and his co-

---

[2]Note that the conclusions indicate that the column headings "with water" and "without water" in Table 1 of the paper by Boudene et al (1977) have been transposed.

Table 4. Comparison of the combustion toxicity of PVC and natural products in the rat under static and dynamic conditions.[a]

Material	Largest Non-lethal Amount (grams)	Estimated LC_{50} (grams)	Zero to 100% Mortality Range (grams)
		Static	
PVC	33	28	33–60
Wool	29	25	29–60
Red Oak	8	13	8–20
Cotton	5	10	5–15
		Dynamic	
PVC	0.5	2.0	0.5–3.5
Wool	0.3	0.4	0.3–0.5
Red Oak	2.5	3.2	2.5–4.1
Cotton	1.5	2.3	1.5–3.1

[a]Cornish et al, 1975.

workers. Particularly, their studies cover a wide range of both natural and synthetic products.

In the studies of Hilado et al (1977), Hilado and Cumming (1978), and Hilado and Huttlinger (1981), materials were combusted by preheating their furnace to 200 °C and then heating the materials at a rate of 40 °C/min. to a maximum of 800 °C. Mice were exposed in a 4.2 l chamber under passive dynamic conditions and the time required to cause incapacitation and death were recorded. In these studies wool caused death more rapidly than PVC or any of the natural materials (Table 6). Of the more than 15 different woods and wood products, all

Table 5. Comparison of the effects of water scrubbing on the combustion toxicity of PVC and douglas fir (Boudene et al, 1977).

Material	Condition	$LC_{50}(g/m^3)$	
		400 °C	800 °C
PVC	Without Water Trap	33.3	25
	With Water Trap	166	145
Douglas Fir	Without Water Trap	37.5	41.6
	With Water Trap	97	83.3

Note: The conclusion in the original paper (Boudene et al, 1977) indicated that the column headings "with water" and "without water" in their Table 1 were transposed. This error has been corrected in the above table.

Table 6. Studies of the relative toxicity on PVC and natural products conducted at the University of San Francisco using mice.

Study	Material	Ti[a] (Min)	Td[a] (Min)
Hilado et al, 1977	Wool[c]	N.D.	6.47-7.60[b]
	Cotton[d]	N.D.	9.66-16.78[b]
	Cotton, flame retarded[e]	N.D.	10.44-20.55[b]
	Hemlock, untreated	N.D.	10.80±0.18
	Hardboard	N.D.	10.86±0.54
	Red oak	N.D.	11.50±0.71
	Particle board, untreated	N.D.	11.82±0.04
	Sisal	N.D.	12.59±3.41
	Douglas fir	N.D	13.62±0.63
	PVC[f]	N.D.	16.37-16.84[b]
Hilado and Cummings, 1978	Wool[c]	5.06-5.45[b]	6.47-7.60[b]
	Cotton[d]	5.90-8.72[b]	9.66-16.78[b]
	Cotton, flame retarded[e]	7.93-8.92[b]	10.44-20.55[b]
	Leather	8.16±0.69	10.22±1.72
	Hemlock, untreated	7 28±2.03	10.80±0.18
	Hardboard, unfinished	8.56±0.37	10.86±0.54
	Red oak[f]	9.09-10.23[b]	11.50-14.50[b]
	Particle board, untreated	9.32±0.39	11.82±0.04
Hilado and Cumming, 1978	Sisal	6.43±2.05	12.59±3.41
	Pigskin	7.16±0.71	12.78±0.77
	Douglas fir[f]	9.84-11.79[b]	13.62-14.76[b]
	Beech	9.69±0.84	13.82±1.69
	Aspen poplar	9.96±1.66	14.06±1.83
	Western hemlock	9.99±0.40	14.37±1.52
	Western red cedar	10.00±0.97	14.91±3.18
	Yellow birch	9.56±0.97	15.09±2.57
	Eastern white pine	10.72±0.85	15.42±0.90
	Southern yellow pine	10.91±0.86	15.56±0.12
	Excelsior	6.60±0.56	15.82±0.11
	Hard board	9.66±2.68	15.90±2.62
	PVC[f]	5.95-12.69[b]	16.37-16.84[b]
Hilado and Huttlinger, 1981	PVC rigid foam	14.15±7.21	24.40±2.83
	Cellulosic board[g]	12.31-16.36[b]	15.90-20.58[b]
	Cellulose insulation, untreated	13.86±1.62	15.85±1 15

[a]mean ± S.D.
[b]expressed as the range of the means for several samples.
[c]2 wool fabric and 1 wool fiber (washed) samples.
[d]2 cotton batting, 1 cotton surgical, 3 cotton fabric samples.
[e]1 cotton batting, 3 cotton fabric flame retarded samples.
[f]2 samples.
[g]6 different types.
N.D. = not determined.

caused death more rapidly than PVC. Only some of the flame retarded cotton materials showed a longer time to death than PVC. No direct correlation was noted between the times to incapacitation (Ti) and death (Td). Numerical rank order for Ti differed quite dramatically in some cases from that of the Td. This was true for sisal, pigskin, excelsior, and one of the PVC samples. Still, the Ti data showed general agreement with the Td data (i.e. that wool performed the worst and that PVC was better than, or comparable to, the other natural products).

Other studies conducted by the Federal Aviation Administration have also reported that many natural products cause incapacitation and death more rapidly than PVC (Smith et al, 1978 and Spurgeon, 1978). Using a system that recirculated combustion gases generated at 600 °C through the furnace and chamber, these studies found that combustion products from flame retarded wool, flame retarded cotton and Douglas fir all required shorter exposure time to cause incapacitation and death than PVC (Table 7).

Considerable research on the combustion toxicity of materials also has been carried out at the University of Pittsburgh. While the actual test procedure has evolved over the years (Kaplan et al, 1983), the test procedure basically has involved the heating of materials at 20 °C/min. In these studies mice have been exposed to various concentrations of combustion gases under dynamic conditions. Each exposure concentration has been produced by diluting the combustion gases as they are fed into a 2.3 1 exposure chamber.

In the studies conducted at the University of Pittsburgh, the relative toxicity of the decomposition products of PVC and natural materials varied, depending on the parameter measured (Table 8). Generally, the data indicated that PVC and PVC-A was more toxic than Douglas fir and SPF wood, but as toxic as cellulose and wool. However, the RD_{50} values, when calculated on a mg/l basis, indicated that PVC was less toxic or at least comparable to Douglas fir.

Table 7. Combustion toxicity evaluations of PVC and natural products conducted by the federal aviation administration.

Study	Material	Ti (min.)	Td (min.)
Smith et al, 1978	PVC	11.91	15.36
	Douglas fir	6.54	8.46
Spurgeon, 1978	PVC foam	5.50	—
	PVC	7.57	14.45
	Wool, flame retarded	2.00	4.17
	Cotton, flame retarded	3.07	4.58

Table 8. Evaluations of the combustion toxicity of PVC and natural products in mice conducted at the University of Pittsburgh.

Study	Material	RD$_{50}$ (mg)	RD$_{50}$ (mg/l)	S1100 (g)	LC$_{50}$ (g)	Asphyxiation Range (g)	LT$_{50}$ (Min)	LTC$_{50}$ (g.min)
Alarie and Anderson, 1979	Douglas fir	34.0		4.0	63.8	55-95		
	PVC-A[a]	16.0		1.0	15.2	8-20		
	Cellulose	N.R.		N.R.	11.9	N.R.		
	Wool	N.R.		N.R.	3.0	N.R.		
Alarie and Anderson, 1981	Douglas fir				63.8		22	1404
	PVC				7.0		10	70
	PVC-A[a]				15.2		15	228
	Cellulose				11.9		21	250
	Wool				3.0		27	81
	SPF wood				48.7		19	925
Anderson et al, 1978	NBS PVC-C[b]		0.25[c]					
	NBS Douglas fir		0.14[d]					
Anderson and Alarie, 1978	PVC-A[a]				13.3[e]		10	
	Douglas fir				63.8[f]		22	
Barrow et al, 1978	Douglas fir		0.24[h]					
	PVC[g]		0.50[i]					
	PVC (plasticized)		0.19[j]					

[a]46% homopolymer: erroneously reported in Alarie and Anderson (1979) as 92% homopolymer.
[b]NBS sample of PVC.
[c]10 minute exposure; 95% confidence limits (0.15-0.42).
[d]10 minute exposure; 95% confidence limits (0.072-0.266).
[e]95% confidence limits (10.5-16.8).
[f]95% confidence limits (58.3-69.9).
[g]no plasticizer or fire retardant.
[h]3 minute exposure; 95% confidence limits (0.15-0.35).
[i]3 minute exposure; 95% confidence limits (0.32-0.78).
[j]3 minute exposure; 95% confidence limits (0.17-0.22).
N.R. = not reported.

More recently, the National Bureau of Standards (1982a) published the results of interlaboratory evaluations of its new test method for evaluating combustion toxicity of materials used in buildings and transportation. In these evaluations, groups of rats were exposed in a 200 l chamber (head only) to the combustion products of PVC and numerous other materials for 30 minutes, followed by a 14 day observation period. A 10 minute exposure period was also used to identify rapid acting toxicants. Materials were combusted at 25 °C below the auto-ignition (non-flaming) and at 25 °C above auto-ignition (flaming). Since the auto-ignition temperature varies depending on the material, 440 °C was also used in order to compare materials at a single temperature. Toxicity was evaluated on the basis of the EC_{50} (concentration causing 50% incapacitation), the LC_{50} with a 14 day post exposure observation period, and the LC_{50} without a post exposure period; any material whose toxicity was more than one order of magnitude lower than wood was considered more toxic than wood (NBS, 1982b).

These interlaboratory evaluations found that the combustion products of PVC had a toxicity similar to that of Douglas fir, red oak, and wool (Table 9). Relative to the natural products, PVC performed the best under non-flaming conditions and at 440 °C. The greatest spread of toxicity values occurred under flaming conditions using a 14 day post exposure period. However, even here the 95% confidence limits indicated that the difference between the LC_{50} (+14 days) values for PVC and these natural products were only slight and well within one order of magnitude.

DISCUSSION

Numerous studies which have been conducted over the last 10 to 20 years have provided us with an extensive body of knowledge about the combustion toxicity of PVC and natural products. Studies conducted by Zikria et al (1977) have shown that wood can produce toxic effects similar to those attributed to PVC, namely incapacitation, pulmonary edema, and immediate or delayed death. Also, Barrow et al (1979) found that both PVC and wood smoke cause upper respiratory tract damage in mice. While we have heard statements that PVC is extremely toxic (Wallace, 1981), the fact is that the published data does not support such a simplistic statement.

At least 16 studies have compared the combustion toxicities of PVC and natural products (Tables 1–9). Since escape and survival time (i.e. the exposure period) is crucial in a real fire, and since exposure time is also an important facet in the measurement of toxicity (Casarett and Doull, 1975), many investigators have compared how quickly materials cause incapacitation and death. Studies by Kishitani (1971), Hilado (1977), Hilado and Cumming (1978), Hilado and Huttlinger (1981), Smith et al (1978), and Spurgeon (1978) have shown that

Table 9. Comparison of EC_{50} and LC_{50} values for PVC and natural materials with rats (NBS, 1982).

Material	EC_{50} 30 min (mg/l)	LC_{50} 30 min + 14 days (mg/l)	LC_{50} 30 min (mg/l)
	Flaming		
Douglas Fir[a]	21.8(15.5-30.7)[c]	41.0(33.0-50.9)	45.0(38.5-52.6)
Douglas Fir[b]	23.5[23.0-24.0][d]	39.8(38.2-41.4)	39.8(38.2-41.4)
PVC	18.5[17.5-19.8][e]	17.3(14.8-20.2)	30.0[f]
Red Oak	34.8(31.1-39.0)	56.8(51.6-62.5)	59.0(54.5-63.9)
Wool	22.3[22.1-22.6][g]	28.2(23.0-34.5)	40.9(38.1-43.8)
	Non-flaming		
Douglas Fir[a]	18.3(14.5-23.0)	20.4(16.4-25.3)	34.8(29.1-41.7)
Douglas Fir[b]	13.5(12.9-14.2)	22.8(20.2-25.8)	29.0(23.4-36.0)
PVC	30.0[h]	20.0(14.7-27.2)	25.0[h]
Red Oak	23[22.5-24.2][c]	30.3(26.0-35.4)	45[f]
Wool	19.7(16.2-24.0)	25.1(22.3-28.3)	29.5(27.8-31.3)
	440°C		
Douglas Fir[a]	—[i]	—[i]	—[i]
Douglas Fir[b]	—[i]	—[i]	—[i]
PVC	30.0[h]	25.0(20.2-31.0)	30.0[f]
Red Oak	—[i]	—[i]	—[i]
Wool	24.5(23.0-26.1)	32.1(30.2-34.1)	35.0(29.0-42.1)

[a]NBS small furnace.
[b]NBS large furnace.
[c]Significantly heterogeneous data.
[d]0-100% affected.
[e]1/6-100% affected.
[f]0% affected.
[g]100% affected.
[h]1/6 affected.
[i]not done when 440°C was equal to or within 50°C of the non-flaming temperature.
( ) = 95% confidence limits.

natural products produced incapacitation and death faster (i.e. require a smaller total dose) than PVC. These findings have been supported by the work of Kishitani and Yusa (1978) where exposure time was narrowly controlled and the amount of material was varied. Here, greater amounts of PVC were required to equal the effects of natural products.

Many other different methods incorporating a fixed exposure period and varying the amount of material combusted have been used for comparing combustion toxicity of PVC and natural products. Studies by Hofmann and Oettel (1969), Cornish et al (1975), Boudene et al (1977), Barrows et al (1978), and the National Bureau of Standards (1982a) have found that PVC is less toxic than, or comparable to many natural products. However, investigators at the University of Pittsburgh (UP) have found that PVC was "more toxic" than wood, but not "much more toxic" (Alarie and Anderson, 1981).

The reasons for the greater relative difference of PVC and Douglas fir in the UP system than in the NBS and other test systems deserves some consideration. In contrast to the NBS protocol, the UP method uses mice rather than rats, a 10 minute post-exposure period instead of 14 days, a much smaller chamber (approximately 100 times smaller than NBS), slow ramping versus fixed temperatures, a dynamic rather than static system, and an external, Lindberg-type furnace as opposed to the internal, Potts (cup) furnace in the NBS test. Differences in species sensitivity or the other factors could have a significant impact on the final results.

Although the greater sensitivity of mice to irritants may account for some of this response, Packham and Crawford (1984) suggest that differences in the combustion and exposure systems are very important. Slow ramping causes combustion products to be produced in a more sequential manner. This means that the exposure period or total dose that the animal experiences will be greater for some chemicals and shorter for others and that the time for possible synergistic or antagonistic interactions will be shorter. The NBS combustion method in comparison, gives a more instantaneous exposure and more equal duration of exposure to all combustion products. Furthermore, the UP method of temperature decreases the toxicity of wood by suppressing char formation which ultimately results in smaller quantities of CO (which are not allowed to collect), producing a shorter exposure period (Packham and Crawford, 1984).

The relative importance of chamber size on the toxic response to PVC combustion products is not clear. The obvious concern here is the effect of chamber size on HCl decay. In the studies by Kishitani (1971) and Hilado et al (1977), the exposure times required to cause death were very similar (15 and 16.37-16.84 minutes, respectively). Although these methods are similar with respect to temperature ramping, test species, and air flow, the test systems differ in chamber size and animal preconditioning to 60 °C for 48 hours. These data suggest that animal preconditioning and chamber size, at least within the range of 4.2 to 56 1 (a factor of 10 difference) has minimal effect on the toxicity of PVC. While it is not possible to know how a larger difference in chamber size (i.e. 100 times larger), as exists between the NBS and UP systems might affect the toxicity, the fact that neither the NBS nor UP test accurately model decay characteristics in a real fire (Packham and Crawford, 1984) may make the question academic.

Although a considerable number of test methods have been developed to evaluate the pyrolysis/combustion toxicity of materials, there presently is no combustion toxicity test method which is predictive of human toxic hazard in a real fire, or which is useful for code setting/regulatory activities (NFPA, 1983; NAS, 1982a; and NAS, 1982b). The reason for this caveat is that combustion toxicology still suffers from the lack of knowledge in three main areas: (1) how to burn the material, (2) how to expose the animals, and (3) whether the rodent is a good model (for man) for all combustion products. While such a statement sounds like an epitaph, the picture actually is not quite that bleak. Information on real world fire conditions has provided some direction in efforts to develop a small scale combustion toxicity test method that is predictive of the potential human response to toxic products generated in a real fire.

Efforts by the National Academy of Sciences, Committee on Fire Toxicology (NAS/CFT) to develop criteria for development of an optimum screening test led to the incorporation of fixed temperature(s) with a short time of combustion (compared to animal exposure time) by the NBS (NAS, 1977 and NBS, 1982a). The choice of a fixed temperature rather than slow ramping here is important because temperatures in a real fire commonly increase at the rate of 1000 °C/min. (Lie, 1974 and Babrouskas, 1981/82). The use of a fixed temperature is appropriate since pyrolysis/combustion is essentially instantaneous and similar to real fires.

Another important area which was addressed by the NAS/CFT is the exposure system. Here the NAS/CFT states that one chamber for both pyrolysis combustion and animal exposure is desirable because it approximates the real fire situation and prevents large loss of combustion particles and gases. Although deaths do occur in the room of fire origin, many fatalities result from smoke that has been transported long distances. In this latter situation, a single chamber may not be predictive of the real toxic potential of a material. O'Mara (1977) points out that such difference in system design may explain some of the differences in the combustion toxicity of PVC. Those systems providing longer transport of gases provide a greater opportunity for "decay" or condensation of HCl from the atmosphere. Clearly, more information is needed on the relationship of real fires to small scale systems for a wide range of materials and on the effect of room (chamber) size and surface area on the transport of combustion gases.

The final part of the combustion toxicity picture remaining to be addressed is the biological system which includes not only the animal, but the biological end-point. Rodents are desirable for screening assays because of their ease of handling and low cost. However, when evaluations go beyond the screening stage to one where the intent is to predict toxicity of a chemical in man, efforts must be made to determine how similar to, or different from man, it is. If more than one chemical is present (i.e. combustion gases), selection of an appropriate animal model and assessing the relative toxicity of materials becomes more difficult. In this case, an animal could be a good model for man with one chemical, but not for another. While this is not a problem when comparing materials which produce the same combustion products, it is a concern when they are different. Presently very little, if any, data is available that addresses this concern. Rodent/primate comparisons would be of value not only to address this issue, but would help in the assessment of biological end-points.

In conclusion, all of the combustion toxicity studies described above show that the relative toxicity of a material can vary depending on many factors such as the temperature of decomposition, the method of decomposition, the air flow, the humidity, the animal species, the environmental temperature, the degree of oxygen depletion, and the treatment of the data. Still, the fact remains that numerous investigators from academia, government, and the private sector have found no evidence that PVC is extremely or unusually toxic relative to natural products. While some test methods indicated that PVC is more toxic than certain natural products, most studies have shown that the combustion toxicity of PVC is less than, or comparable with that of natural products. Considering that large amounts of combustibles in homes and that multi-story buildings are a fact of life, the data does not support contentions that PVC is unusually toxic or hazardous.

REFERENCES

Alarie, Y.C., and Anderson, R.C., "Toxicologic and Acute Lethal Hazard Evaluation of Thermal Decomposition Products of Synthetic and Natural Polymers," *Toxicol. Appl. Pharmacol.*, 51:341–362 (1979).

Alarie, Y.C., and Anderson, R.C., "Toxicologic Classification of Thermal Decomposition Products of Synthetic and Natural Polymers," *Toxicol. Appl. Pharmacol.*, 57:181–188 (1981).

Anderson, R.C., and Alarie, Y.C., "Approaches to the Evaluation of the Toxicity of Decomposition Products of Polymeric Materials Under Thermal Stress," *J. Combust. Toxicol.*, 5:214–221 (1978).

Anderson, R.C., Stock, M.F., and Alarie, Y.C., "Toxicologic Evaluation of Thermal Decomposition Products of Synthetic Cellular Materials," Part 1, *J. Combust. Toxicol.*, 5:111–129 (1978).

Babrouskas, V., "Will the Second Item Ignite?" *Fire Safety J.*, 4:281–292 (1981/82).

Barrow, C.S., Alarie, Y.C., and Stock, M.F., "Sensory Irritation and Incapacitation Evoked by Thermal Decomposition Products of Polymers and Comparisons with Known Sensory Irritants," *Arch. Env. Health*, 33: 79–88 (1978).

Barrow, C.S., Lucia, H., Stock, M.F., and Alarie, Y.C., "Development of Methodologies to Assess the Relative Hazards from Thermal Decomposition Products of Polymeric Materials," *Am. Ind. Hygiene Assoc. J.*, 40: 408–423 (1979).

Birky, M.M., Halpin, B.M., Caplan, Y.H., Fisher, R.S., McAllister, J.M., and Dixon, A.M., "Fire Fatality Study," *Fire and Materials*, 3:211–217 (1979).

Boudene, C., Jouany, J.M., and Truhart, R., "Protective Effect of Water

Against Toxicity of Pyrolysis and Combustion Products of Wood and Poly (Vinyl Chloride)," *J. Macromol. Sci-Chem.*, All(8):1529-1545 (1979).

Casarett, L.J., Toxicologic Evaluation in "Toxicology, the Basic Science of Poisons," L.J. Casarett and J. Doull, editors, MacMillan Publishing Company, Inc., New York, p. 13 (1975).

Committee on Fire Toxicology. "Fire Toxicology: Methods for Evaluation of Toxicity of Pyrolysis and Combustion Products," Report No. 2, National Academy of Sciences, Washington, D.C. (August 1977).

Cornish, H.H., Hahn, K.J., and Barth, M.L., "Experimental Toxicology of Pyrolysis and Combustion Hazards," *Environ. Health Perspect.*, 11:191-196 (1975).

Darmer, Jr., K.I., Kinhead, E.R., and DiPasquali, L.C., "Acute Toxicity in Rats and Mice Exposed to Hydrogen Chloride Gas and Aerosols," *Amer. Ind. Hyg. Assoc. J.*, 35:623 (1974).

Fassett, D.W., XL. Organic acids, anhydrides, lactones, acid halides, and amides, thioacids, in "Industrial Hygiene and Toxicology," 2nd edition, Patty, F.A., editor, Interscience Publishers, New York, pp. 1962-1972 (1963).

Further Development of a Test Method for the Assessment of the Acute Inhalation Toxicity of Combustion Products. NBSIR 82-2532, U.S. Department of Commerce, National Bureau of Standards, National Engineering Laboratory, Center for Fire Research, Washington, D.C. (June 1982(a)).

Halpin, B.M., and Berl, W.G., "Human Fatalities from Unwanted Fires," Report NBS-GCR-79-168, National Bureau of Standards, Washington, D.C. (1978).

Harland, W.A., and Wooley, W.D., "Fire Fatality Study-University of Glasgow," Information Paper IP 18/79, Building Research Establishment, Glasgow, Scotland (1979).

Henderson, Y., and Haggard, H.W., *Noxious Gases*, 2nd. ed., Rinehold, New York, p. 126 and 128 (1943).

Hilado, C.J., and Cumming, H.J., "Relative Toxicity of Pyrolysis Gases From Materials: Effects of Chemical Composition and Test Conditions," *Fire and Materials*, 2(2):68-79 (1978).

Hilado, C.J., and Huttlinger, P.A., "Toxic Hazards of Common Materials," *Fire Tech.*, pp. 117-182 (August 1981).

Hilado, C.J., Lopez, M.T., and Damant, G.H., "Relative Toxicity of Pyrolysis Products from Some Upholstery Fabrics," *J. Coated Fabrics*, 6:155-175 (1977).

Hofmann, H.T., and Oettel, H., "Comparative Toxicity of Thermal Decomposition Products," *Mod. Plastics*, 46(101):94-97 (1969).

Hofmann, H.T., and Sand, H., "Further Investigations Into the Relative Toxicity of Decomposition Products Given Off from Smoldering Plastics," *JFF/Combust. Toxicol.*, 1:250-258 (1974).

Kaplan, H.L., Grand, A.F., and Hartzell, G.E., "Combustion Toxicology: Principles and Test Methods," Technomic Publishing Co., Lancaster, PA.

Kimmerle, M.G., "Aspects and Methodology for the Evaluation of Toxicological Parameters During Fire Exposure," *J. of Fire and Flammability Combustion Toxicology (Suppl.)*, 1:4-56 (1974).

Kishitani, K., "Study on Injurious Properties of Combustion Products of Building Materials at Initial Stage of Fire," *J. Faculty Engineer*, U. Tokyo, *31(1)*:1–35 (1971).

Kishitani, K., and Yusa, S., "Study on Evaluation of Relative Toxicities of Combustion Products of Various Materials," *J. Faculty Engineer*, U. Tokyo, *35(1)*:1–17 (1979).

Lie, T.T., "Characteristic Temperature Curves for Various Fire Severities," *Fire Technol.*, *10(4)*:315–326 (1974).

O'Mara, M.M., "Combustion of PVC," *Pure Appl. Chem.*, *49*:649–660 (1977).

Packham, S.C., and Crawford, M.B., "An Evaluation of Smoke Toxicity and Toxic Hazard of Electrical Non-Metallic Tubing Combustion Products," submitted to *J. Fire Sci.* (1984).

Packham, S.C., and Hartzell, G.E., "Fundamentals of Combustion Toxicology in Fire Hazard Assessment," *J. Testing Evaluation*, *9*:341–347 (1981).

Report of the Committee on the Toxicity of the Products of Combustion to the Standards Council of the National Fire Protection Association, *Fire Journal*, *77(2)*:21–27 (1983). Chairman Joseph E. Johnson.

Smith, P.W., Crane, C.R., Sanders, D., Abbott, J. and Endecott, B., "Material Toxicology Evaluation by Direct Animal Exposure," Aviation Toxicology Laboratory, DOT, FAA Civil Aeromedical Institute, Oklahoma City, Okla. (1978).

Spurgeon, J.C., "The Correlation of Animal Response Data with Yields of Selected Thermal Decomposition Products for Typical Aircraft Interior Materials," National Aviation Facilities Experimental Center, Federal Aviation Administration, prepared for the U.S. DOT, Report No. FAA-RD-78-131 (1978).

Steinhagen, W.H., Swenberg, J.A., and Barrow, C.S., "Acute Inhalation Toxicity and Sensory Irritation of Dimethylamine," *Amer. Ind. Hyg. Assoc. J.* *43(6)*:411–417.

"TLVs Threshold Limit Values for Chemical Substances and Physical Agents in the Work Environment with Intended Changes for 1982," ISBN: 0-936712-39-2, American Conference of Governmental Industrial Hygienists, Cincinnati, OH, pp. 9 and 21.

Wallace, D.N., "Dangers of Polyvinyl Chloride Wire Insulation Decomposition, 1. Long Term Health Impairments: Studies of Firefighters of the 1975 New York Telephone Fire and of Survivors of the 1977 Beverly Hills Supper Club Fire," *J. Comb. Toxicol*, *8*:205–232 (1981).

Workshop on Combustion Product Toxicity Summary of Presentations," NBSIR 82-2634, U.S. Department of Commerce, National Bureau of Standards, National Engineering Laboratory, Center for Fire Research, Washington, D.C. (September 1982(b)).

Yuill, C.H., "Physiological Effects of Products of Combustion," *Amer. Soc. Saf. Eng. J.*, *19*:36–42 (1974).

Zikria, B.A., Ferrer, J.N., and Flock, H.F., "The Chemical Factors Contributing to Pulmonary Damage in 'Smoke Poisoning'," *Surgery*, *71*:704–709 (1972).

An Evaluation of Smoke Toxicity and Toxic Hazard of Electrical Nonmetallic Tubing Combustion Products

S. C. PACKHAM

BETR Sciences, Inc.
996 South 1500 East
Salt Lake City, Utah 84105

M. B. CRAWFORD

Crawford Consulting, Inc.
125 Sunrock Drive
Folsom, California 95630

ABSTRACT

Small-scale smoke toxicity tests were performed on polyvinylchloride-based electrical nonmetallic tubing (PVC/ENT) using the National Bureau of Standards (NBS) protocol for nonflaming combustion. An LC_{50} of 28.5 mg/L ($\pm$ 9.25) was determined, placing PVC/ENT smoke in a toxicity category comparable to smoke from wood.

Four large-scale tests were conducted in a 21,734-L room (8 × 12 × 8 ft). In tests involving 18 to 30 linear in. of PVC/ENT with and without fuel-contributing heat sources, it was learned that: a) 30 in. of PVC/ENT developed a 3.3-mg/L smoke concentration and an insufficient HCl concentration to cause death or significant symptoms in test animals (rats) and b) in the presence of a relatively small fire (5-lb wood crib), heat and carbon monoxide became life-limiting prior to HCl. In additional tests involving approximately 60 in. of PVC/ENT degraded under 2.5 Watts/cm² heat flux, a smoke concentration of 7.6 mg/L was developed. Animal fatalities in these tests were shown to be from heat stress rather than smoke inhalation. The measurable percents of theoretical yield of HCl in small-scale tests ranged from 64 to 96 percent and in large-scale tests from 23 to 50 percent. The reasons for these differences are discussed.

Research Funded by: Carlon, An Indian Head Company, Three Commerce Park Square, 23200 Chagrin Boulevard, Cleveland, Ohio 44122.

INTRODUCTION

POLYVINYLCHLORIDE-BASED ELECTRICAL NONMETALLIC tubing (PVC/ENT) is used to contain and protect electrical wiring. PVC/ENT has a considerable history of use in Europe with no reported instances of initiating fire or contributing to fire hazard, but its introduction into United States' markets more recently has prompted controversy [1,2]. Most of the debate has revolved around the toxicity of gaseous combustion products from PVC/ENT and their potential contribution to hazard in real fires.

From the early 1960's to 1980 ENT, along with other PVC conduit, gained from 9 to 54 percent of the market share, replacing metal conduit [1]. These marketing inroads were likely due in part to reduced construction costs related to its easier installation and its elimination of ground fault electrical fire hazards such as those responsible for the MGM Grand Hotel fire in which 85 persons lost their lives [3].

The positive features of PVC/ENT, both of installation and insulation, are due to its inherent organic make-up. PVC, its basic component, is a combustible polymer known to release hydrogen chloride (HCl) and carbon monoxide (CO). Since PVC/ENT applications would replace noncombustible metal with combustible PVC, there has been concern over its possible contribution to toxic hazard, vis-a-vis, smoke inhalation. For example, some have speculated that based on LC_{50}'s developed in small-scale toxicity tests, that only 18 to 30 in. of pyrolyzed PVC/ENT in an average room would be " . . . sufficient to create a dangerous situation and kill humans in 10 to 15 minutes . . ." [4].

The small-scale toxicity test method used in the hazard projection noted above was developed at the University of Pittsburgh [5,6]. Since the relative toxicity, or LC_{50}, of PVC/ENT tested according to the National Bureau of Standards (NBS) protocol [7] was not available in the literature, this study included such a determination. In addition, respiration rate was clinically monitored during these tests. To evaluate the accuracy of the hazard projection quoted above and to further understand the toxic hazard of PVC/ENT in full-scale configurations, a series of tests were conducted in an $8 \times 12 \times 8$-ft room.

METHODS

Materials

The PVC/ENT compound used in this study contained by weight:

PVC homopolymer	87.78%
Inert noncombustible inorganic material	4.39%
Stabilizers and lubricants	7.83%

Typical wiring insulation compounds similar to that used in this study contain by weight:

PVC homopolymer	58%
Inert noncombustible inorganic material	12%
Plasticizers and other additives	30%

In this series of tests, the weight of PVC/ENT was 5.083 g/in. The weight of the PVC insulation on the three conductors was 0.55 g/in. The weight of the three copper wires (stripped of insulation) was 2.235 g/in. Thus, the combined weight of the PVC/ENT and the PVC wiring insulation (excluding the bare copper wire) was 5.633 g/in. The PVC/ENT comprised 90.24 percent of that weight, with the wiring insulation being 9.65 percent.

It has been found by prior research [8] that PVC homopolymer contains 58 percent by weight of unburnable HCl, 38 percent of burnable hydrocarbons and 4 percent of carbonaceous char.

Theoretically then, during combustion, PVC/ENT could release 50.91 percent of this weight ($0.8778 \times 0.58 = 0.5091$) as HCl gas; 41.19 percent as other volatile materials ($[0.878 \times 0.38] + 0.0783 = 0.4119$); leaving 4.39 percent as inert material; and 3.51 percent as char ($0.8778 \times 0.04 = 0.0351$).

Similarly, one could estimate that, during combustion, the wire insulation could release 33.64 percent of its weight ($0.58 \times 0.58 = 0.3364$) as HCl gas; 52.04 percent as other volatile materials ($[0.58 \times 0.38] + 0.30 = 0.5204$), leaving 12 percent as inert material; and 2.32 percent ($0.58 \times 0.04 = 0.0232$) as char.

Analytical Procedures for HCl

Determinations of HCl concentrations in the 200-L chamber and the full-scale test room from thermal degradation of PVC/ENT were made using an ion-selective electrode technique.

During room tests, discrete 50-cc grab samples were drawn at 1, 3, 5, 6, 7, 8, 9, 10, 12, 15, 20, 25 and 30 minutes through four 40-in. × 6-mm (I.D.) teflon tubes supported in stainless-steel tubes wrapped with electrical resistance heating tape to prevent condensation. Intra-sample line temperatures were maintained at 110°C ± 5°. Orientations of sample lines are shown in Figure 1; two being located 5.5 ft above the floor and two at the 3.5-ft level. The probe lines terminated approximately 2 in. from each of four animal retaining cages.

An electrical pump having a 1.1-L/min capacity was used to draw test room air through the sample lines for 10 seconds (Tests 1 and 2) or 1 second (Test 3) to purge them of nonrepresentative atmospheres

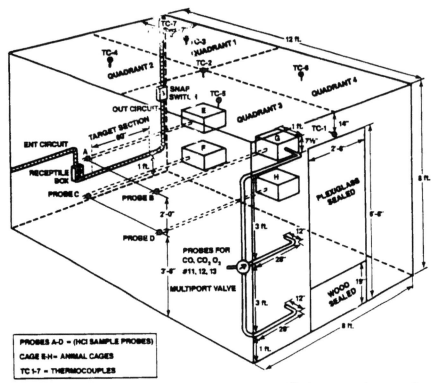

Figure 1. Test room configuration of wall-mounted ENT, thermocouples, sampling probes and animal cages used in tests 1 and 2.

prior to actual collection of the 50-cc sample volume. In the final room test (Test 4), room-air samples were taken using both a 1-second and a 10-second draw time as well as by a glove-box technique adapted to the room (see Figure 2). The three techniques revealed no difference in HCl capture or retention.

Individual sample volumes were drawn into plastic syringes and capped. A measured aliquot of a solution of known concentration of HCl was added to the syringe and agitated for at least 20 seconds to permit complete absorption of gaseous HCl. Total Cl⁻ content in the solution was then measured using an ion-selective electrode/meter system which was calibrated to a series of solutions of known CL⁻ concentrations in moles/L. The millivolt response of the ion-selective electrode/meter system was linearly related to Cl⁻ concentrations above 5×10^{-5} moles/L. Parts per million (ppm) of HCl was then calculated using the following formula:

$$\text{HCl, ppm} = ([\text{Cl}^-]_f = [\text{Cl}^-]_i) \left(\frac{V_s}{V_g}\right) (2.466 \times 10^{10})$$

where: $[Cl^-]_f$ = final chloride in concentration in moles per liter after washing the room gas with the HCl solution.

$[Cl^-]_i$ = initial chloride ion concentration in moles per liter of the HCl absorbing solution

V_s = volume of the HCl absorbing solution

V_g = volume of the gas sample

Based on conditions of a 50-cc sample dissolved into 2 cc of 1×10^{-3} M Cl absorbing solution and on assumptions of meter readability (sensitivity) of either 0.1 mg or 0.3 mg, an estimated accuracy table of the technique was developed (Table 1). Although some titration methods are recognized as being more precise, the ion-selective method was the method of choice in these studies since its accuracy was sufficient for the levels of HCl anticipated and it allowed an accumulation of many data points throughout the 30-minute experiments and avoided dependence on time-averaged values.

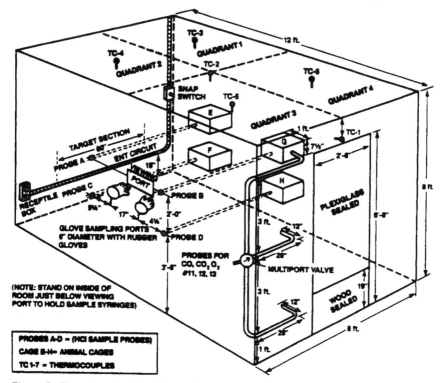

Figure 2. Test room configuration of wall-mounted ENT, thermocouples, sampling probes with glove-box modification and animal cages used in tests 3 and 4.

Table 1. Estimated repeatability of the analytical for HCl method. *

PPM	@0.1 MV	Repeatability in PPM @0.3 MV
4 (detection limit)	± 4	—
12 (detection limit)	—	± 12
40	± 4	± 12
80	± 4	± 12
450	± 5	± 16
1150	± 13	± 30
8900	± 37	± 110

* Based on readability specifications of the equipment.

Small-Scale Toxicity Test Procedures

Samples of 1-in. O.D. PVC/ENT were tested using the NBS test method [7]. The test samples were light blue in color and had corrugated or ribbed surfaces. Individual test specimens were cut from stock and weighed before being placed in the 9-cm (I.D.) × 15-cm deep quartz beaker of the cup furnace. Net mass loss was determined by weighing the residual char following each test.

All test procedures, animal exposure, care and handling as well as the monitoring of exposure chamber oxygen (O_2), carbon monoxide (CO) and carbon dioxide (CO_2) were carried out as specified by the NBS protocol [7]. An auto-ignition temperature of 650°C was established in preliminary tests on PVC/ENT without animals. The auto-ignition temperature of pelletized PVC samples used in the NBS interlaboratory evaluation ranged from 600 to 625°C [7]. The higher auto-ignition temperature represented by PVC/ENT may reflect either a difference in formulation or physical configuration or both.

According to the NBS protocol, nonflaming combustion tests are conducted at a furnace temperature of 25°C below the auto-ignition temperature. Therefore, the current tests of PVC/ENT were at a furnace setting of 625°C. To start each test, a single piece of weighed PVC/ENT was placed in the pre-heated furnace, the door to the 200-L exposure chamber was closed and a 30-minute head-only exposure of six rats was begun. Male laboratory rats were purchased from Simonson Laboratories in Gilroy, California and were of the Fisher 344 strain, weighing between 225 and 300 g at the time of testing. A series of tests was conducted in which sample masses loaded into the furnace yielded a range of 0 to 100-percent lethality and a statistical LC_{50} was calculated.

Small-Scale Toxicity Test Results

In Table 2 initial sample specimen weights, smoke concentrations, lethality/exposure ratios and maximum percent carboxyhemoglobin (COHb) values determined in the seven small-scale tests are reported. Smoke concentrations were calculated by dividing the amount of sample placed in the furnace by 200 L; this being a slightly more liberal estimate of dose than concentration determinations based on mass of sample lost as thermal degradation products. Approximately 95 percent of the original mass was lost from sample specimens as volatiles, pyrolyzates and combustion products in each case.

An LC_{50} of 28.5 mg/L was obtained from these data using the Finney method. The 95-percent confidence limits for this LC_{50} were 18.7 mg/L and 37.2 mg/L. All deaths occurred post exposure. The maximum COHb percentage observed at the end of the 30-minute tests was 25.1; the lowest 12.1. There was no correlation observed between percent COHb and smoke concentration. This non-correlation likely reflects the overwhelming influence of irregular respiration rates (Figure 3) on the accumulation of COHb given relatively low ambient CO concentrations (937-1250 ppm). Clinical observations of the animals were made. Corneal opacity along with a frothy discharge from the nose and mouth were noted upon removal of the animals from the test exposure. These signs increased in severity commensurate with smoke concentrations. Deaths were delayed with progressive body weight loss being typical prior to death.

Ambient concentrations of HCl in the 200-L chamber were char-

Table 2. Results of NBS toxicity test.

Conducted for PVC/ENT at 625°C Non-Flaming				
Original Specimen W (g)	Smoke Concentration* (mg/)	Lethalite		Max COHb (%)
		Exposure	Post-Exposure	
2.51	12.6	0/6**	0/5**	22 5
3.44	17.2	0/6	0/5	13 8
4.15	20.8	0/6	1/5	15 9
5.40	27.0	0/6	2/5	25 1
6.03	30.2	0/6	3/5	22 2
6.84	34.2	0/6	3/5	21 5
8 00	40.0	0/6	5/5	12.1

LC_{50} = 28.7 mg/ ; 95% confidence limits = 18.7 to 37 2 mg/
*Calculated based on mass of sample placed in the furnace
**One of the six animals exposed in each test was sacrificed for COHb determination leaving 5 for post-exposure observation

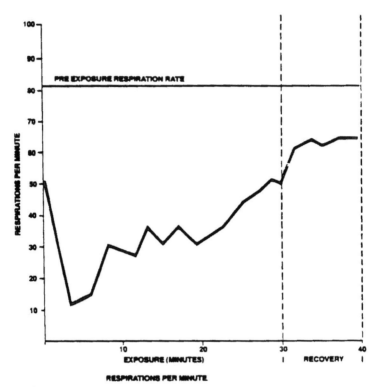

Figure 3. Respiration rate of rats exposed to smoke from ENT for 30 minutes in a 200-L chamber.

acterized for smoke concentrations of 3.7 and 8.5 mg/L (Figure 4). These experiments, conducted in the absence of animals, produced maximum concentrations of 1150 ppm and of 2200 ppm HCl, respectively, from these particular PVC/ENT loadings. The theoretical yields of HCl from the 0.78-g and 1.69-g samples used to develop these smoke concentrations can be estimated to be 1200 ppm and 2950 ppm, respectively, assuming 58-percent yield for HCl from the constituent PVC homopolymer (homopolymer being 87.78 percent of the PVC/ENT by weight). Thus, 96 percent and 75 percent of these theoretical yields were detected. Average HCl concentrations over the 30 minutes following placement of the PVC/ENT specimens in the cup furnace fell to 900 ppm and 1800 ppm in these respective tests. There was a greater decay factor observed in the higher smoke concentration test.

Full-Scale Toxic Hazard Test Procedures

Four tests were conducted in a 21,734-L room. Wall mounted 3/4-in. (O.D.) PVC/ENT carrying three lines of #12 THHN conductor in-

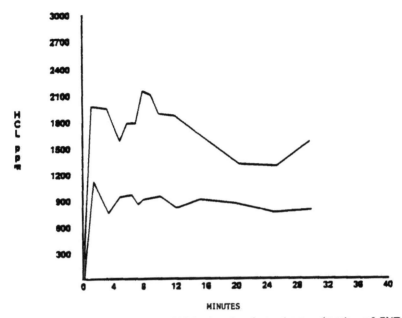

Figure 4. HCl concentrations in a 200-L chamber from the combustion of ENT at 625°C. Upper curve from a 1.69-g sample (8.2 mg/L) and the lower curve from a 0.73-g sample (3.7 mg/L).

sulated with flexible PVC was exposed to preselected fire test conditions. Two configurations were used. The one shown in Figure 1 was used for tests involving thermal degradation of sections nominally 18 in. (Test 1) and 30 in. (Test 2) in length. The configuration illustrated in Figure 2 was used in tests designed to expose up to 60 in. of the PVC/ENT wiring assembly (Tests 3 and 4). Wall and ceiling surfaces in all tests were unpainted 1/2-in. gypsum wallboard mounted on 24-in. centerline spaced 2 × 4-in. fir studs.

These test configurations represented atypical applications in that PVC/ENT is most frequently installed behind wallboard. It is reasoned that surface-mounted PVC/ENT would represent an absolutely worst-case scenario for a toxic hazard evaluation.

Bio-assay verification of toxic hazard was sought using eight rats per test held in four cages (i.e., 2 per cage). Two cages were positioned 5.5 ft above the floor; the other two were directly below at the 3.5-ft level. The cages, constructed of 1 × 2-in. wooden frames and non-metallic screening, were suspended on light chains from the ceiling. Specifications of strain, age, sex and body weights for rat subjects used in those tests were identical to those used in small-scale tests described above and they were obtained through the same supplier.

Sampling lines for CO, CO_2, and O_2 were located adjacent to the door at the 1-ft, 4-ft and the 7-ft level above the floor. They extended 1 ft into the room and were joined to a common line outside the room via a multiple-port valve before being led to analytical instrumentation. Batch sample determinations for all three gases from each of the in-room probe lines were made every 90 seconds. CO and CO_2 determinations were made using non-dispersive infrared instrumentation. Paramagnetic instrumentation was used for O_2 concentration determinations.

Temperature within the test room was monitored from 7 thermocouple (TC) leads. TC1 was located 14 in. below the ceiling directly above the doorway. TC's 2 through 7 each extended 2 in. below the ceiling with TC2 being located in the geometric center of the room and TC's 3, 4, 5 and 6 being located at the geometric centers of room quadrants 1, 2, 3, and 4, respectively, as illustrated in Figures 1 and 2. The seventh thermocouple (TC7) was positioned 2 in. below the ceiling near the back right-hand corner of the room assuming an observational orientation at the door facing into the room.

Percent light transmission across a 1-meter vertical path extending upward from a point 3 ft. above the floor was measured using light source/photocell apparatus. Signals from chemical CO, CO_2 and O_2, thermal and light transmission instrumentation were fed on-line to a Tektronix 4051 computer for immediate storage.

Full-Scale Toxic-Hazard Test Results

Tests 1 and 2. In Test 1, quartz lamps were calibrated and positioned to cast a nominal 2.5-Watts/cm^2 incident heat flux on approximately 30 linear in. of the PVC/ENT wire assembly. This "nonfuel" energy application was contrasted in Test 2 with a burning 5-lb wood crib, a "fuel-contributing" energy source. The wood crib in Test 2 was positioned to also project a nominal 2.5-Watts/cm^2 heat flux on the target section of PVC/ENT. Both tests lasted 30 minutes.

At the conclusion of Test 1, char and discolored residue extended over 27 in. of the PVC/ENT target section. Paper on the gypsum wallboard was burned away 8 to 10 in. above the PVC/ENT and the exposed gypsum surface was well calcined. Deformed, but otherwise uncharred ENT extended about 1.5 in. beyond both ends of the charred section showing minimal involvement of ENT beyond the high energy region.

Of the 212.44 g original weight of the 27-in. section, 139.94 g remained; indicating a net weight loss of 72.50 g. Bare wire contributed 60.35 g to the 27-in. burned section, meaning that 79.59 g or 52 percent of the combustible ENT available was not volitalized or combusted in the test. Forty grams of this remaining mass was black char and 39.59 g was partially decomposed. ENT surfaces against the gypsum showed only partial charring and decomposition, evidence of a distinct wall effect.

The 72.50-g mass loss from ENT in Test 1 represented a 3.3-mg/L smoke concentration in the closed and sealed 21,734-L room. Test room concentrations of HCl rose to 750 ppm in 9 minutes and decayed to 600 ppm by the end of the test (Figure 5). Light transmission decreased to 50 percent by 6 minutes, to 20 percent by 12 minutes, and to a low of 15 percent by 18 minutes. Thereafter, light transmission increased slightly and was 18 percent at 30 minutes (Figure 6). Room temperature, averaging all thermocouple readings, increased steadily from 22 °C at the start of the test to 48 °C at 18 minutes and to 52 °C at the end of the 30-minute test (Figure 7). Percent CO_2 reached a maximum level of 0.04 percent and CO concentration did not exceed 0.02 percent during the test. No detectable drop in O_2 was measured.

Animals, removed at the end of the test, showed signs of moderate inflammation of the naries and slight to moderate discharge of a clear exudate from the nose and mouth. Eyes of the test animals were clear following exposure and showed no signs of cornial opacity, residual irritation or alterations in sensitivity to blink or cornial reflexes. One animal sacrificed for blood was found to have 26.7-percent COHb. No deaths were observed among the remaining animals during or follow-

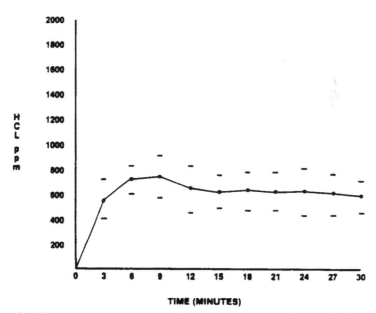

TIME (MINUTES)

Figure 5. HCl measured in test room from combustion of 30 in. of ENT using a 2.5-watt/cm² nonfuel radiant energy source.

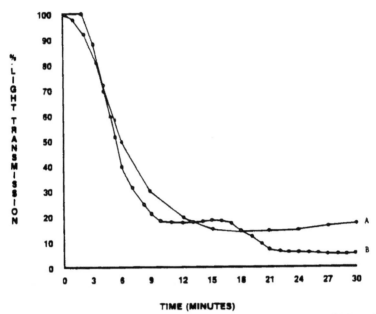

TIME (MINUTES)

Figure 6. Percent light transmission vs. time in the test room when 30 in. of ENT were burned using a 2.5 watts/cm² nonfuel radiant energy source (curve A) and a flaming wood crib as a radiant energy source (curve B).

ing the test, and 14-day body-weight records showed full recovery and growth trends.

In Test 2, in which a 5-lb wood crib ignited with 100 mL of ethanol was used as a "fuel-contributing" energy source, 62.7 g was lost from 17.5 in. of the PVC/ENT wire assembly. The pre-burn weight of 17.5 in. of PVC/ENT assembly can be calculated to be 137.69 g [(5.083 + 2.785) × 17.5 = 137.69 g]. The bare wire portion of that initial weight was estimated to be 39.11 g (2.235 g/in. × 17.5 = 39.11 g) leaving 98.58 g as the initial weight of ENT and insulation combustibles. Therefore, 64 percent of the original mass was lost as volatiles and combustion products. Of the 35.88 g remaining, 21.47 g was char and 14.41 g was deformed and partially decomposed. A mass loss of 62.7 would normally represent a smoke concentration 2.9 mg/L, but there was a considerable contribution from the wood crib. The HCl concentrations measured in the test room during Test 2 increased to 410 ppm in 9 minutes (Figure 8). After a slight decline to 350 ppm at 15 minutes, there was a second increase to 440 ppm at 21 minutes. From 21 minutes to the end of the test a steady decay was observed; the final HCl concentration being 190 ppm.

Light transmission decreased to 40 percent by 6 minutes and 22 percent at 9 minutes, where it remained until about 16 minutes into the test, at which point it decreased steadily to 6 percent by 27 minutes (Figure 6). This irregular curve reflects the manner in which the wood crib burned. The ceiling room temperature recorded during Test 2 (Figure 7) shows an aggressive fire early in the first 3 minutes of the test (room temperature reading 125 °C) followed by a regressive phase which plateaued at about 6 minutes. At this point the O_2 level in the test room had also dropped to 18.5 percent. From 9 to 15 minutes into the test the wood crib again burned vigorously and ceiling temperatures reached a maximum level of 130 °C. At 15 minutes O_2 in the test room fell below 16 percent which markedly and perceptably altered the burning behavior of the crib.

None of the animals survived Test 2. Blood COHb was found to be less than 15 percent even though room concentrations of CO were over 1500 ppm at 20 minutes into the test. This relatively low COHb percentage eliminates CO as a probable cause of death. HCl concentrations can be ruled out as a significant factor in the cause of death in Test 2 since all animals survived Test 1 in which HCl levels were higher. Physiological heat stress remains the more likely, though unsubstantiated, cause of death in Test 2.

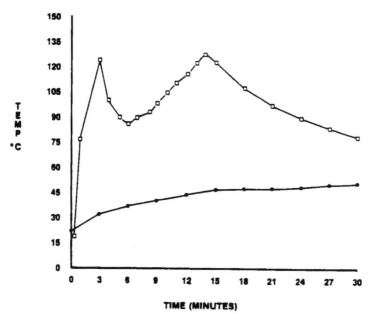

TIME (MINUTES)

Figure 7. Temperatures recorded in the test room when 30 in. of ENT were burned using a 2.5 watts/cm² nonfuel radiant energy source (closed circles) and a flaming wood crib as a radiant energy source (open squares).

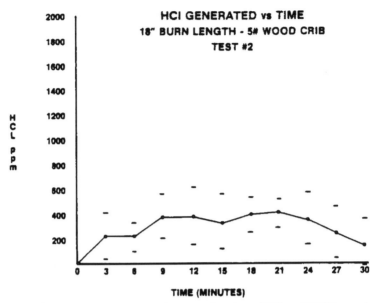

Figure 8. HCl measured in test room from combustion of 18 in. of ENT burned with a flaming 5-lb wood crib.

Test 3 and 4. In Test 3, 2.5 Watts/cm² were applied to about 60 in. of the ENT circuit (Figure 2). Post-test observation confirmed a 59-in. effected section. The pretest calculated weight of combustible material in 59 in. of the PVC/ENT assembly was 332.34 g. Almost 50 percent of that mass, 166.99 g, was lost as volatiles and combustion products. There was 105.8 g of char remaining (31.85 percent of original mass) and 59.55 g, or 17.9 percent, was only partially charred.

HCl concentrations (Figure 9) increased to 400 ppm in 6 minutes and peaked at 540 ppm at 9 to 12 minutes into the test. By the end of the test only 375 ppm HCl remained. Light transmission decreased to 50 percent by 8 minutes, to 20 percent by 15 minutes and remained near that level throughout the remainder of the test (Figure 10). The room temperature increased gradually to a 72 °C maximum at the termination of the test. Only 0.17-percent CO_2 was recorded and CO reached a maximum level of 400 ppm. There was essentially no O_2 depletion.

The four test animals housed in the two rear cages, nearest the heat source, were found dead at the end of the test. The remaining four animals emerged alive with symptoms similar to animals of Test 1. These four animals survived a 14-day post-test observation period during which time body weights and general health improved following a 2 to 3-day depression.

Results from Test 3 raised several questions. Since more than twice

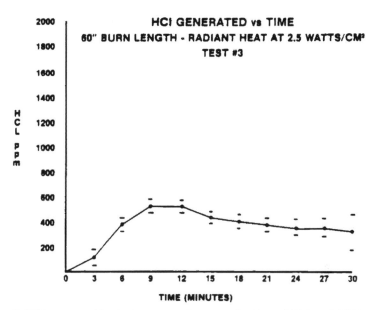

Figure 9. HCl measured in the test room from combustion of 60 in. of ENT with a 2.5-watts/cm² nonfuel radiant energy source.

the mass of ENT was lost as in Test 1, it was surprising that a higher HCl level was not observed. A change from a 10-second draw time through HCl probes in Tests 1 and 2 to a 1-second draw time in Test 3 was hypothesized as an explanation. A fourth test, a repeat of Test 3, was therefore undertaken during which both 1-second and 10-second draw times were implemented. In addition, a glove-box arrangement was installed so that results from direct sampling could be compared with those from probe-line sampling. Lastly, since only animals in the cages near the test wall died during Test 3, it was suspected that heat-stress was the cause. In Test 4, two test animals were anesthetized, instrumented with rectal thermocouples, and placed in cages at the 5.5-ft level. All other test conditions in Test 4 were identical to Test 3.

A total of 150.55 g was lost from 59 linear in. of PVC/ENT in Test 4. HCl concentration peaked at 445 ppm at 12 minutes into the test (Figure 11). Points and range bars plotted in Figure 11 represent all data points taken using the three sampling procedures. Room temperatures in Test 4 paralleled those in Test 3 (Figure 12). Percent light transmission profiles were also nearly identical. Recordings from animal rectal probes are plotted in Figure 13. The animal nearer the heat lamps died and developed a 47 °C terminal body temperature. The other animal survived a body core temperature of 41 °C was measured by the end of the test.

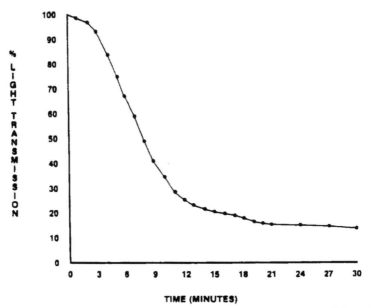

Figure 10. Percent light transmission vs. time in the test room when 60 in. of ENT were burned using a 2.5-watts/cm² nonfuel radiant energy source.

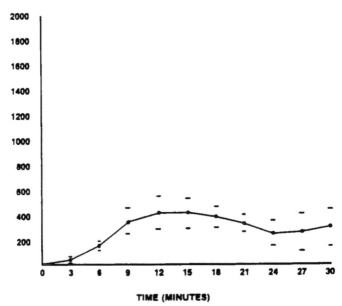

Figure 11. HCl measured in the test room from combustion of 60 in. of ENT using a 2.5-watts/cm² nonfuel radiant energy source.

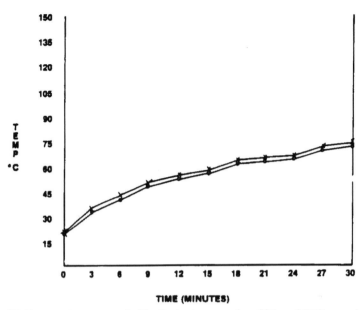

Figure 12. Temperatures recorded in the test room when 60 in. of ENT were burned using a 2.5-watts/cm³ nonfuel radiant energy source (test 3—X's, test 4—closed circles).

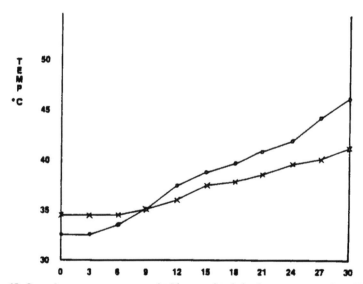

Figure 13. Rectal temperatures recorded from animals in the test room when 60 in. of ENT were burned using a 2.5-watts/cm³ nonfuel radiant energy source (closed circles—data from animal nearer the heat sources; X's—data from animal nearer the door).

ROOM TEMPERATURE vs TIME

• TEST #3 - 60" BURN LENGTH - RADIANT HEAT AT 2.5 WATTS/CM²

x TEST #4 - 60" BURN LENGTH - RADIANT HEAT AT 2.5 WATTS/CM²

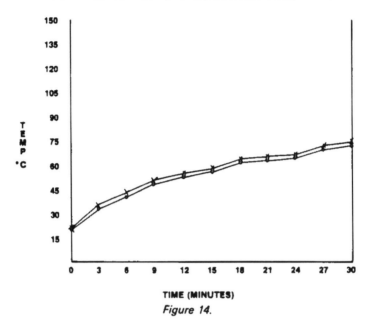

TIME (MINUTES)

Figure 14.

DISCUSSIONS AND CONCLUSIONS

The toxicity of smoke from PVC/ENT samples was determined using the NBS nonflaming test protocol. An LC_{50} of 28 mg/L having 95-percent confidence limits of 18.7 to 37.2 mg/L was obtained. Smoke from Douglas fir, tested according to the same protocol has a reported LC_{50} of 22.8 mg/L with 95-percent confidence limits ranging from 13.4 to 32.2 mg/L [7]. The acute toxicity of PVC/ENT and wood as evaluated using the NBS protocol are, therefore, comparable. In contrast to these results, the Pittsburgh toxicity test method reports PVC to be 4.2 to 9.1 times more toxic than Douglas fir [6].

Differences seen in the relative toxicity of PVC compared to wood in these two test methods are most likely a function of the manner in which the two tests are conducted rather than to a difference between rats (NBS test) and mice (Pittsburgh test). In the NBS test, the sample specimen is brought quickly to a temperature just below its auto-ignition. Combustion products are released quickly and their concentrations are held relatively constant in the animals' exposure chamber

ANIMAL CORE - BODY TEMPERATURE vs TIME
RECTAL PROBE INSTRUMENTATION
• CAGE A ANIMAL - TOP CAGE NEAREST HEAT SOURCE
x CAGE B ANIMAL - TOP CAGE NEAREST DOOR
TEST #4

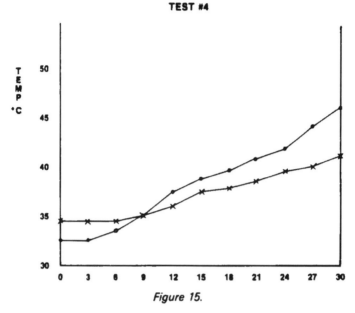

Figure 15.

for virtually the entire 30 minutes of the test (Figure 4). Samples in the Pittsburgh test are heated by a furnace programmed to increase temperatures at a rate of 20 °C/min. Combustion products are usually released within discrete temperature windows and are passed by the animals at a flow rate of 20 L/min. These two procedures present much different exposures to the test animals. In the NBS test, concentrations of combustion products are generally lower (being diluted in the 200-L chamber) and are held for a longer (30-minute) exposure. In the Pittsburgh test very high combustion product concentrations are often seen since considerably less dilution occurs prior to presentation to the animal, but, the residence times of smoke in the animal exposure system of the Pittsburgh test are of shorter duration. Exposure times of less than 10 minutes are not uncommon.

For corrosive toxicants such as acid gases, high concentrations can cause lesions and edema of the respiratory tract and lungs from rather brief exposures with delayed clinical disfunction and death being a likely consequence. These effects may be relatively independent of the duration of the exposure once an effective concentration threshold is surpassed. In contrast, CO produces no tissue lesions. If the animal survives a brief, but high exposure concentration, full recovery may be expected. On the basis of exposure dynamics alone one might predict that a greater amount of wood (a CO generator) than PVC (an acid gas generator) would be needed to cause lethality in the Pittsburgh test.

Apart from these differences in exposure dynamics between the two tests, there is yet another possible reason why wood smoke appears more toxic in the NBS test. The NBS test, in contrast to the Pittsburgh test which increases sample temperature at a rate of 20 °C/min., is designed to hold the sample specimen just below its auto-ignition point. For wood in the NBS test, this means that a char is quickly formed and large quantities of CO are produced throughout the 30-minute test. Under the programmed rising temperature regimen of the Pittsburgh test, wood samples reside in a pre-flaming state for a relatively short time and the CO produced is not allowed to collect in the animal exposure system. Once flaming occurs, much less CO is released per unit mass of sample burned. Therefore, in the NBS test more CO, the primary toxicant in wood smoke, is produced per unit mass of material and the CO is retained in the static 200-L chamber, effecting a longer exposure time. The Pittsburgh test, in contrast, produces less CO from a given quantity of wood and the flow-through transport of combustion products limits the animal's exposure time. The relative toxicity of PVC smoke to wood smoke thus appears to change when comparing results of the NBS and Pittsburgh tests.

A second objective of the research reported here was to investigate toxic hazard presented by PVC/ENT in a real fire in an average size room. Since HCl is recognized as a significant toxicological component of PVC smoke, room tests were conducted to investigate the degree to which HCl yields from PVT/ENT in small-scale tests could be used to estimate HCl concentrations when burned in full-scale configurations. About 51 percent by weight of PVC/ENT (i.e., rigid, PVC) can theoretically be released as HCl. At atmospheric pressures and ambient temperatures recorded during the laboratory and full-scale tests conducted above one mole of HCl (1 mole = 36 g) would be expected to occupy 24.2 L. Given these assumptions, calculations can be made of the percent of theoretical yield of HCl measured in specific test runs. In the Pittsburgh test [4] 5.4 g of ENT produced an average of 6,377 ppm of HCl over 10 minutes. Since a flow rate of 20 L/min is used in the Pittsburgh procedure, one can estimate a 6,377-ppm HCl concentration in an effective volume of 200 L. The theoretical yield of HCl from 5.4 g ENT would be 2.75 g, or 0.08 moles, or 1.94 L. In an effective volume of 200 L, a 1.94-L volume represents roughly a 1-percent V/V ratio or an estimated theoretical yield of 10,000 ppm HCl. In the Pittsburgh test, then, approximately 64 percent (i.e., 6,377 ppm divided by 10,000 × 100) of the theoretical yield was detected. By comparison, in the NBS tests reported above, 75 to 96 percent of the theoretical yield was measured in the 200-L chamber.

In Test 1 of the full-scale test series, 72.50 g of PVC/ENT was volatilized. Assuming 51 percent of this mass evolves as HCl, one would estimate a theoretical HCl yield of about 1500 ppm. A maximum of 750 ppm was measured in Test 1 or about 50 percent of the theoretical yield. In Test 2, 34 percent of PVC/ENT's potential yield of HCl was detected. In Tests 3 and 4, only 23 and 25 percent of the theoretical yield of HCl were measured.

These calculations suggest that both the Pittsburgh and the NBS tests may over estimate the effective exposure concentrations of HCl which may be expected in room size tests. HCl concentrations seen in full-scale tests were substantially below the theoretical yield. This discrepancy is probably a function of a combination of factors including variable atmospheric conditions (relative humidity) and the particular sample/wall configuration which existed in Tests 1 through 4.

Neither the NBS nor the Pittsburgh small-scale tests accurately modeled decay characteristics of HCl in real-fire conditions simulated in Test 2 above. A comparison of Test 1 and Test 2 appears to suggest that hazardous levels of CO and heat in an actual fire may develop life-threatening conditions at least as early, if not before, the evolution of hazardous HCl concentrations.

Results from these tests failed to verify the toxic hazard projection quoted in the introduction. Even though HCl in the test room reached 750 ppm, the symptoms exhibited by the animals were minimal. The lack of severe clinical signs at these levels of HCl may seem somewhat puzzling in light of Henderson and Haggard's published "dangerous for even short exposure" levels of 1,000 to 2,000 ppm [9]. The basis for the Henderson-Haggard table of physiological responses to various concentrations of HCl is a series of references dealing mostly with regulations and industrial codes [10–13]. A more recent authority, the 1980 Registry of Toxic Effects of Chemical Substances [14], based on emperical data gives 660 ppm as the lower lethal concentration of HCl for man in a 2 hour exposure. Our observations tend to confirm this more current view and cast some doubt on the validity of conventional wisdom based on earlier sources. They further suggest that toxic hazard, when extrapolated from small-scale test results should be verified and confirmed by full-scale fire simulations to: a) substantiate the toxicological extrapolations and to b) establish the relative significance of toxic hazard to other fire hazards of heat and smoke density.

Since PVC/ENT is usually installed behind the wall, it would seem highly unlikely that its combustion products could add significantly to the toxic hazard of smoke from a fire initiating inside the room. This study clearly did not address the transport of smoke or HCl to remote locations. An investigation of hazards developed in multi-room configurations is, therefore, suggested as a fruitful area for future research.

ACKNOWLEDGEMENT

The authors are pleased to acknowledge the contributions of Dr. Gregory Smith of B. F. Goodrich Chemical Company for assisting in the design and evaluation of HCl analysis techniques and to Mr. D. L. Kent, consultant, for assistance in data reduction and graphics.

We are also indebted to the staff of the Weyerhaeuser Fire Technology Section of Longview, Washington; namely, Mr. R. R. McNeil, manager, Mr. J. White, Dr. J. Shaw, Mr. C. Bailey and Mr. J. Luebbers. Also, Mr. H. Stacy, presently at Southwest Research Institute and formerly of Weyerhaeuser, who was responsible for laboratory management during the NBS Toxicity Test Runs.

This research was funded by Carlon, An Indian Head Company, Mr. H. F. van der Voort, project monitor.

REFERENCES

1. Flax, S., "The Dubious War on Plastic Pipe," *Fortune* (February 7, 1983).
2. Rawls, R., "Fire Hazards of Plastic's Spark Heated Debate," *Chem. and Engineering News* (January 3, 1983).
3. Best, R., and Demers, D. P., "Investigation Report on the MGM Grand Hotel Fire, Las Vegas, Nevada, November 21, 1980," National Fire Protection Association.
4. Alaire, Y. C., "Toxicity of Smoke Emitted From 4 Commercial Samples When Heated at 20¼C/Minutes and Attempts to Project the Hazard in Humans," An unpublished "Final Report."
5. Alaire, Y. C., and Anderson, R. C., "Toxicological Classification of Thermal Decomposition Products of Synthetic and Natural Polymers," *Toxicology and Applied Pharmacology*, Vol. 57, No. 2, pp. 181-188 (February 1981).
6. Anderson, R. C., and Alaire, Y. C., "Approaches to the Evaluation of the Toxicity of Decomposition Products of Polymeric Materials Under Thermal Stress," *Journal of Combustion Toxicology*, Vol. 5, pp. 214-221 (May 1978).
7. "Further Development of a Test Method for the Assessment of the Acute Inhalation Toxicity of Combustion Products," NBSIR 82-2532, U.S. Department of Commerce, National Bureau of Standards, National Engineering Center for Fire Research, Washington, D. C. 20234 (June 1982).
8. O'Mara, M., "A Comparison of Combustion Products Obtained From Various Synthetic Polymers," *Journal of Fire and Flammability/Combustion Toxicology* (August 1974).
9. Henderson, Y., and Haggard, H., "Noxious Gases and the Principles of Respiration Influencing the Action," American Chemical Society Monograph Series (Second Edition). Reinhold Publishing Corporation, 330 West 42nd Street, New York. USA (1943).
10. State of California, Dept. Ind. Relations, Minutes Indust. Accident Comm., 1939.

11. Dalla Valle: U.S. Public Health Service, "Principles of Exhaust Hood Design," 1939.
12. Bowditch, Drinker, Drinker, Haggard and Hamilton: *J. Ind. Hyg. Toxicol.*, 22, 251 (1940).
13. State of Connecticut, Sanitary Code, Regulation 281, 1936.
14. Registry of Toxic Effects of Chemical Substances, 1980.

—————————— 8 ——————————

A Bioassay Model for Testing the Incapacitating Effects of Exposure to Combustion Product Atmospheres Using Cynamolgus Monkeys

DAVID A. PURSER

Department of Inhalation Toxicology
Huntingdon Research Centre plc
Huntingdon, Cambs., U.K.

ABSTRACT

A bioassay model has been developed to test the time course and degree of incapacitation produced by exposures to thermal decomposition products from polymeric materials. A battery of physiological tests was used during separate exposures to atmospheres of carbon monoxide, hydrogen cyanide, hypoxia, hypercapnia and heated air. Each atmosphere was designed to simulate one aspect of the conditions commonly encountered in fires. Measurements were made of the animals' respiration, cardiac function and respiratory blood gases. Neurological function was monitored by measurements of the electroencephalogram auditory cortical evoked potentials and peripheral nerve conduction velocity. Hypoxia (10% oxygen) caused muscle weakness, a decrease in nerve conduction velocity, abnormal cardiac function accompanied by a fall in blood pressure and central nervous system depression. At 1000 ppm carbon monoxide, venous carboxyhaemoglobin levels reached 30%. There was a reduction of nerve conduction velocity and in some cases severe central nervous system depression. At 60 ppm hydrogen cyanide had a slight depressive effect on the central nervous system, while at 80–150 ppm severe central nervous system depression and incapacitation occurred. The main result of 5% carbon dioxide exposure was a three-fold increase in respiratory minute volume. It is concluded that the model is capable of detecting early physiological signs of incapacitation induced by fire conditions. It is suggested that exposures to a combination of hydrogen cyanide and carbon monoxide with accompanying changes in cerebral blood flow during attempts to escape from fires may be a cause of collapse and subsequent death.

INTRODUCTION

PROBABLY THE MOST important single piece of information to emerge from the reasonably detailed fire statistics collected in the United Kingdom over the last 30 years is that there was an approximately four-fold increase in the numbers of fatal and non-fatal casualties reported as overcome by toxic gas and smoke between 1955 and 1971 [1]. This increase has continued over the last decade [2], and occurred despite the fact that the numbers of burn casualties have decreased. The total annual numbers of fires have remained approximately constant over this period of time.

All fire atmospheres contain potentially lethal products, but a point that emerged from the statistics is the importance of the time course and degree of incapacitation occurring during the early stages of exposure. A large percentage (60% for dwellings in 1980) [2] of fatal casualties are found in the room of origin of the fire, often close to an exit, where they become incapacitated during attempts to escape.

A previous report presented in this Journal [3] summarised a five year programme of work during which an animal model was used to study incapacitating effects during sublethal exposures to atmospheres of thermal decomposition products from polymeric materials decomposed under a range of temperatures and oxygen environments [3,4].

The aim of each experiment was to find an atmosphere concentration which would produce signs of incapacitation in cynomolgus monkeys within a 30 minute exposure period. This was achieved by carrying out preliminary range finding studies in which animals were exposed to atmosphere concentrations increasing in separate exposures from very low concentrations to concentrations where early signs of a physiological response were seen. The animals were then exposed at this atmosphere concentration and the physiological responses were measured using a battery of tests designed to detect incapacitating changes in the state of the animals.

Before this work was carried out it was necessary to establish the validity of the model and to obtain background data on individual gases commonly found in fires. For the work reported here cynomolgus monkeys were exposed for 30 minute periods to hypoxia, hypercapnia, warm air and air mixtures of carbon monoxide and hydrogen cyanide. The physiological tests fell into 2 categories:

1. Standard tests used to monitor the general state of the animal (measurements of respiration, cardiac function and blood gas exchange).
2. Special tests designed to measure the integrity of the animal's nervous system and detect changes in its level of consciousness (EEG), central nervous processing (auditory cortical evoked response) and neuromuscular function (peripheral nerve conduction velocity).

The EEG was measured because it provides a good indication of the degree of consciousness of the animal and the general integrity of its central nervous system. The auditory cortical evoked response results from complex central nervous system activity involving transmission across many synapses and it is known that hypoxia can affect synaptic delay [5]. Changes in this parameter may reveal early effects on the central nervous system. Nerve conduction velocity was measured because it has been reported that CO can reduce the velocity of impulse conduction in peripheral nerve [6].

METHODS

Two groups of 4 cynomolgus monkeys were exposed to atmospheres of single gases for 30 minute periods as follows:

Group 1 (Animals numbers 267, 269, 109 and 111)

Carbon monoxide	1000 ppm	balance air
Heat	40 °C	balance nitrogen

Group 2 (Animals numbers 259, 261, 263 and 265)

Hypoxia	15% and 10% O_2	balance nitrogen
Hypercapnia	5% CO_2	balance air
Hydrogen cyanide	60 ppm	balance air

The experiments on CO and HCN were later extended to cover higher concentrations (CO 1000–8000 ppm) [7] (HCN 80–150 ppm) [8].

Since it was not practical to monitor all the physiological parameters on an animal during a single exposure, each animal was exposed to each atmosphere on 3 occasions. This enabled the full range of parameters to be measured at least once on each animal for each atmosphere.

The animal under test was placed in a chair and the test atmosphere was supplied to the animal's snout via a face mask.

Exposures to hypoxia, hypercapnia and warm air were carried out using a 70 litre rectangular chamber. An airflow of 10 l/min was then established through the chamber for head only exposures. For exposure to CO and HCN a different method of exposure was used. The gas/air mixture was supplied to the face mask via a 4-way adaptor. The adaptor was also connected to a 3 litre anaesthetic bag, a sampling line and an extract line. This produced a leak-proof system from which the animal could be supplied a test atmosphere via a pneumotachograph.

Each test session consisted of 3 consecutive 30 minute periods:

1. Pre-exposure period, during which baseline parameters were established.
2. Exposure period, during which the effects of test atmospheres were examined.

3. Recovery period, during which the parameters under examination could be observed as they returned to pre-exposure levels.

The following parameters were measured during the pre-exposure, exposure and recovery periods:

ATMOSPHERE CONCENTRATION—For atmospheres consisting of CO_2, CO or low O_2 the concentration of test gas was measured continuously, CO and CO_2 by infra-red analysers and O_2 by a paramagnetic analyzer. HCN atmospheres were set up at approximately 60 ppm HCN by dilution of a standard gas mixture and the concentration monitored by intermittent measurements using colorimetric tubes. For later exposures at 80–150 ppm [8] cyanide levels were measured by silver nitrate titration [9] from samples taken in 0.1M sodium hydroxide solution.

PHYSIOLOGICAL PARAMETERS—Air flow into and out of the lungs was measured by means of a pneumotachograph connected to a differential gas pressure transducer. Electrical integration of the flow signal gave the animal's tidal volume (TV), and from this could be calculated the respiratory minute volume (RMV) and respiratory rate (RR).

Blood gas measurements were made on 3 occasions for all test atmospheres except HCN. Blood samples were taken from a catheter placed in the femoral artery, and analysed on a Radiometer blood gas analyser. Measurements were made of arterial PO_2, PCO_2 and pH. During CO exposures the percentage of carboxyhaemoglobin (% COHb) was estimated by the method of Commins and Lawther [10].

Cardiac effects induced by the test atmospheres were assessed by measurements of the ECG and the arterial blood pressure.

Neuromuscular conduction was measured by means of a muscle twitch test. The ulnar nerve was stimulated electrically at the elbow and wrist by 0.2 ms, 16 v square wave pulses repeated once per second. The electrical activity produced by contraction of the hypothenar muscle was detected by means of electrodes placed in the skin of the hand. Delays in neuromuscular conduction were detected by increases in response latency (the time interval between stimulus and response), and nerve conduction time by differences in latency between elbow and wrist stimulation. The nerve conduction velocity could then be calculated by dividing the distance between the elbow and wrist by the nerve conduction time.

The auditory cortical evoked potential (EP) was recorded from electrodes implanted on the surface of the auditory cortex. The responses to 0.5 ms square wave clicks repeated once per second were averaged over 255 presentations using a Neurolog signal averager. Measurements were made of changes in the latency and amplitude of various components of the response during the test session.

The electroencephalogram (EEG) was recorded throughout the test session from electrodes implanted into the animal's skull resting on the dural surface.

Channel 1—left occipital to left temporal
Channel 2—left temporal to left frontal
Channel 3— right occipital to right frontal

Measurements were made of changes in the power spectrum of the EEG by integrating the activity in each of the 4 main EEG wave bands (delta 1-4 Hz, theta 4-8 Hz, alpha 8-13 Hz and beta 13-30 Hz). In general central nervous system depression is indicated by increases in low frequency (delta and theta) activity and, to some extent, a decrease in the beta activity. Central nervous system activation is represented by increases in beta activity.

RESULTS

The effects of exposures to test atmospheres on physiological parameters are shown as follows:

Table 1— Mean respiratory minute volumes during test sessions.

Table 2— Percentage change in arterial blood gas levels during the exposure period relative to the mean of the pre-exposure and recovery periods.

Table 3— Percentage change in nerve conduction time between elbow and wrist, and latency of response to elbow stimulation during exposure period relative to the mean of the pre-exposure and recovery periods.

Table 4— Percentage change in distribution of EEG power between the 4 major frequency bands.

Changes resulting from exposures to the test atmospheres were detected in all the physiological parameters measured, but in most cases the changes were not dramatic and it seems that for most exposures the effects were at the threshold level of functional impairment.

Carbon Monoxide

Ten of the twelve exposures to 1000 ppm CO seemed to have little general effect upon the animals beyond a slight relaxing influence reflected in a small progressive decrease in RMV and PO_2 towards the end of the exposure period. No effects were seen on the EEG and no consistent effects on the EP. The effects upon nerve conduction velocity were somewhat variable since one animal showed no change while another showed a slight effect (17% decrease) and another a moderate effect (47% decrease).

The fourth animal (no. 269) was severely affected by CO on two occasions showing a severe respiratory depression towards the end of the

Table 1. Mean respiratory minute volumes during test sessions (ml).

Atmosphere	Pre-exposure		Exposure		Recovery		% change[x]
		SD		SD		SD	
10% O_2	1681	(448)	2261	(736)	1566	(406)	+ 39
15% O_2	2206	(905)	1981	(564)	1625	(231)	+ 9
5% CO_2	1499	(245)	4227	(832)	1541	(379)	+ 178
40°C heat	1617	(308)	1412	(149)	1469	(273)	− 8
1000 ppm CO	1996	(557)	1574*	(690)	1548	(395)	− 14
60 ppm HCN	1652	(287)	2136*	(765)	1627	(410)	+ 39

[x]% change during exposure period relative to mean of pre-exposure and recovery period.
*RMV during last 5 minutes of exposure period.
SD   Standard deviation.

exposures and requiring resuscitation on one occasion. The animal became unconscious after 20 minutes of exposure on both occasions showing a marked (90%) decrease in nerve conduction velocity (Figure 1) and severe EEG changes consisting of a large increase in slow delta and theta wave activity and a decrease in alpha and beta activity (Figure 2).

Table 2. % change in arterial blood gas levels during the exposure period relative to the mean of the pre-exposure and recovery periods.

PO_2

Animal no.	15% O_2	10% O_2	CO_2	CO
267	− 44	− 67	+ 12	nm
263	− 33	− 62	+ 18	− 6
269	− 40	− 63	+ 27	− 8
Mean	− 39	− 64	+ 17	− 7

nm − not measured

The % change in the PCO_2 caused by each test atmosphere relative to the control periods were as follows:

PCO_2

Animal no.	15% O_2	10% O_2	CO_2	CO
267	0	− 10	+ 23	nm
263	− 10	− 22	+ 25	− 20
269	− 3	− 17	+ 44	+ 5
Mean	− 4	− 16	+ 31	(− 8)

Table 3. *% change in nerve conduction time between elbow and wrist, and latency of response to elbow stimulation during exposure period relative to the mean of the pre-exposure and recovery periods.*

Atmosphere	% change in conduction time elbow to wrist	% change in elbow latency
10% O₂	+ 9	+ 12
15% O₂	0	0
CO₂	− 2	+ 2
Heat	− 9	+ 1.5
CO	+ 39	+ 14
HCN	+ 4.5	+ 3.5

On the other occasion when this animal was exposed to carbon monoxide and recordings were made of the EP, the effects of exposure were mild and no consistent changes in the EP were observed.

The uptake of CO and its incorporation into carboxyhaemoglobin was measured in three animals. The venous carboxyhaemoglobin levels increased linearly throughout exposure reaching 28, 28 and 30% respectively by the end of the exposure period (Figure 3). The half-life for the excretion of CO from the blood was approximately one hour. On one occasion both arterial and venous carboxyhaemoglobin levels were measured. The arterial level reached 34% by the end of the exposure period while the venous level was only 28%.

Table 4. *% change in distribution of EEG power between the 4 major frequency bands.*

Gas	Animal No.	delta 1-4 Hz	theta 4-8 Hz	alpha 8-13 Hz	beta 13-30 Hz	Total 1-30 Hz
10% O₂	263	+ 85	+ 47	+ 16	+ 9	+ 40
CO₂	261	+ 18	− 8	− 4	− 19	− 7
	263	+ 10	− 1	0	− 16	0
Heat	111	− 9	+ 7	0	− 5	− 4
CO	111	+ 4	+ 5	+ 11	0	+ 5
	109	− 12	+ 1	+ 14	+ 4	+ 1
	259	− 8	− 20	− 10	− 12	− 12
	269	+ 121	+ 113	+ 88	− 26	+ 60
HCN	109	+ 53	+ 43	− 2	− 12	+ 19
	259	+ 81	+ 22	− 9	− 14	+ 15

The table shows the % increase in activity in each waveband during the exposure period relative to the mean of the pre-exposure and recovery periods.

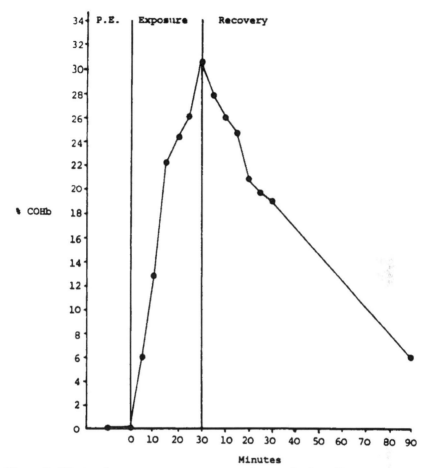

Figure 1. Effects of exposures on neuromuscular conduction. The upper graph shows the latency of the muscle twitch following stimulation at the elbow and may reflect effects on either nerve or muscle. The lower graph shows the time taken for conduction along the nerve between elbow and wrist and reflects effects on the nerve alone.

Heat

Exposure of the animals to 40 °C for 30 minutes caused no obvious changes in any parameter beyond a slight possible increase in nerve conduction velocity.

Hypoxia

Of the atmospheres tested, 10% oxygen produced the most general and consistent effects on the parameters measured. The slight

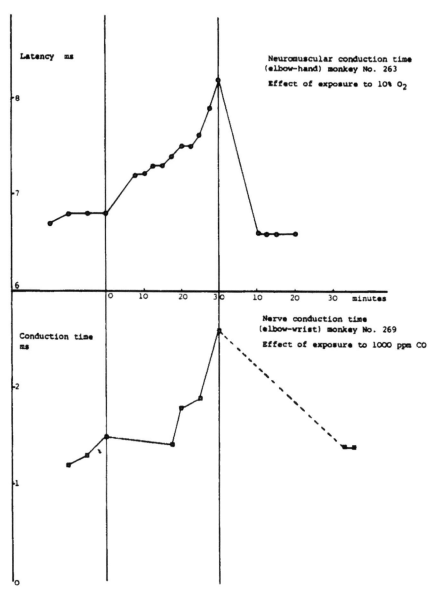

Figure 2. Effect of an exposure to 1000 ppm CO upon the EEG (monkey no. 269). As with HCN there is a massive increase in slow wave activity and a decrease in fast activity as the monkey becomes semiconscious, but the change occurs later in the exposure.

hyperventilation was insufficient to prevent a 64% reduction in arterial PO_2, and the venous PO_2 fell to as little as 10 mmHg. Although animals under these conditions appeared to be superficially normal and were definitely conscious, there was clear evidence of muscle weakness and impairment of reflexes. The blood pressure was reduced (−18%) and there was a small increase in heart rate (+6%). Perhaps more significantly, 3 out of the 4 animals tested suffered ECG abnormalities during exposure consisting of paroxysmal premature ventricular extrasystoles.

There was also a fairly consistent but slight (10%) increase in the response latency following peripheral nerve stimulation in the 4 animals studied, which tended to increase towards the end of the exposure period (Figure 1). Ten percent oxygen also affected the central nervous system as shown by the effects on the EP and EEG. The auditory cortical evoked response consisted of two major potential peaks, an early sharply defined peak with a latency of approximately 17 ms, and a large amplitude, long duration delayed response with a variable peak latency in the 50-70 ms range. Ten percent oxygen caused a slight (7%) increase in the latency of the primary and late peaks, and a 20-40% reduction in the amplitude of the late response. Clear signs of CNS depression consisting of increases in EEG delta and theta activity and decreases in alpha and beta activity were also observed.

In contrast to the above, exposure to 15% oxygen appeared to have no effect on the animals beyond a 39% reduction in the arterial PO_2 and a slight increase in heart rate.

Carbon Dioxide

The most significant feature of exposures to 5% CO_2 was a large (approximately three-fold) increase in RMV resulting from an immediate increase in both respiratory rate and tidal volume. This hyperventilation caused a 17% increase in PO_2 and there was a 31% increase in PCO_2 and a slight respiratory acidosis (mean arterial pH 7.213). The cardiovascular effects of CO_2 exposure appeared to be minimal although one animal suffered a period of cardiac arrhythmia during exposure. There were no effects upon the peripheral nerve conduction velocity but there did appear to be some degree of central nervous system depression, consisting of an increase in slow wave EEG activity, and EP changes consisting of a 7% increase in latency and a 20% reduction in amplitude of both primary and late responses.

Hydrogen Cyanide

Although HCN exposures at approximately 60 ppm caused some effects on most of the parameters measured, the effects were small. There was a slight increase in RMV, but no effects were seen on cardiovascular parameters (blood gases were not measured during HCN exposures). There was no effect upon peripheral neuromuscular conduction. There was, however, clear EEG evidence for some degree of central nervous system depression occurring at the end of the exposure periods, and although the latency of the EP was unaffected there was a 26% reduction in the amplitude of the late response.

However, during later experiments in the 80-150 ppm HCN range [8] the effects of exposure were much more dramatic and a different pattern of response occurred. At some time during the exposure period a large increase in respiration occurred, the RMV increasing by up to 4 times. Within 1-5 minutes of the start of this hyperventilatory episode the animals became semiconscious, the respiration slowed and RMV decreased, a pattern of slow deep breaths developing with a pause at the end of expiration between each successive breath. This was accompanied by severe effects upon ECG rate and rhythm and severe central nervous system depression (Figure 4).

DISCUSSION

Although the physiological changes recorded in these experiments were in most cases not dramatic they provided a validation of the model in that effects were detected in all the parameters measured, even though it seems that for most exposures we were at the threshold for functional impairment. This provided a sound basis for the more severe effects recorded in later experiments at higher gas concentrations and during smoke exposures.

Apart from providing essential background data for the later primate smoke exposures the experiments also revealed some likely effects of individual gases on fire victims.

Hypoxia

These experiments showed that although 15% oxygen caused a 40% reduction in PO_2 there was no significant functional impairment of the albeit sedentary animals, while the further reduction to 10% oxygen produced almost immediate clear signs of impairment in all the parameters measured. Although the animals appeared to be superficially normal and were definitely conscious it is very likely that their state of alertness was impaired and there were signs of muscle weakness and impairment of reflexes.

Carbon Monoxide

In contrast to the consistent effects of low oxygen exposure, the effects of CO were much more variable, despite the fact that the 3 animals showed very similar patterns of CO uptake.

Under the present experimental conditions, with the animals sitting quietly in restraining chairs, it seems that the animals were able to function normally, although peripheral nerve conduction velocity was impaired in 3 out of the 4 animals tested. However, from other experience of the effects of CO containing atmospheres on cynomolgus monkeys in this laboratory [3] it appears that at levels of 30-40% COHb this apparently normal situation can deteriorate very suddenly particularly if a previously quiet animal becomes briefly active, and an animal can pass from apparent normality to collapse in a few minutes. This effect is clearly demonstrated by the response of one animal (no. 269) in the present study which collapsed on 2 of the 3 occasions during which it was exposed to CO, showing severe central nervous system depression and a 90% decrease in nerve conduction velocity. However, during a third exposure the effects were extremely mild and no catastrophe collapse point occurred during exposure.

In active, free-moving animals exposed in more recent experiments [11] a similar collapse towards the end of exposure was observed at lower carboxyhaemoglobin levels. It therefore appears that an important feature of the effects of CO relevant to the fire situation is that there is little incapacitative effect until the COHb concentration has increased to a minimum blood level, after which the occurrence of a sudden crisis with collapse depends on precipitating factors such as the degree of activity of the subject.

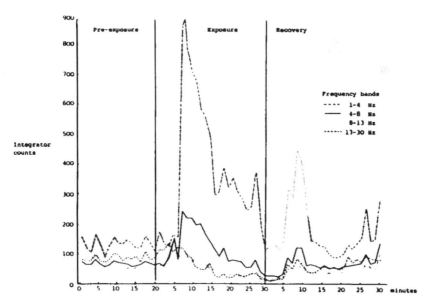

Figure 3. Percentage of total haemoglobin as carboxyhaemoglobin in venous blood for monkey no. 111 exposed to 1000 ppm CO for 30 minutes.

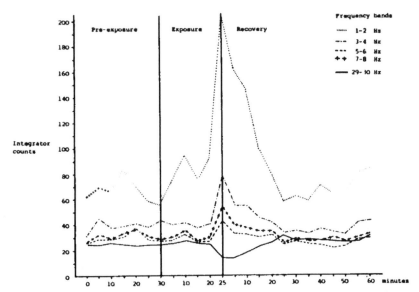

Figure 4. Effect of an exposure to 147 ppm HCN gas upon the EEG (monkey no. 825). A massive increase in slow wave (delta) activity is followed by a decrease in fast (beta) activity as the monkey passes into a state of semiconsciousness.

Carbon Dioxide

The most significant feature of exposure to 5% CO_2 was not so much a toxic effect of the gas itself as the importance of the massive increase in RMV caused by it. This would result in a much more rapid uptake of any other toxic gases present in a fire atmosphere such as CO or HCN.

However, in addition to the slightly incapacitating effects of the hyperventilation there were some signs of slight cardiovascular effects and slight cerebral depression. In man the first symptoms occur above 3% CO_2 and signs of intoxication have been reported after 30 minutes exposure at 5%, while 7-10% produces unconsciousness within a few minutes [12].

Hydrogen Cyanide

Unlike CO where the occurrence of incapacitation depends upon a dose accumulated over a period of time, incapacitation during hydrogen cyanide exposure appears to be more critically dependent upon the concentration to which the victim is exposed. Thus over a 30 minute period the effects are minor at concentrations up to approximately 80 ppm. Over a narrow range between 80-180 ppm dramatic signs of incapacitation occurred at some stage during a 30 minute exposure following a short episode of marked hyperventilation, and over this range there was a loose relationship between atmosphere concentration and time to incapacitation [8]. At concentrations of above approximately 180 ppm HCN a severe episode of hyperventilation occurs within 1-2 minutes of the start of exposure followed by a loss of consciousness within 5 minutes.

It therefore seems that concentrations of HCN in fires below 60 ppm are not likely to produce a significant impairment of escape capability in fire victims, but that at concentrations in the 100-200 ppm range it may be a factor in producing a rapid "knock down" effect on victims causing them to remain in the fire and later die as a result of the accumulation of CO or other factors.

Mechanisms of Incapacitation Resulting from Atmospheres Producing Tissue Hypoxia

The effective tissue hypoxia induced by the three test atmospheres (10% O_2, CO and HCN) might be expected to affect the performance of an animal in 3 ways:

1. Cerebral hypoxia may have a range of effects from clouding of judgment, through serious confusion and loss of co-ordination, to unconsciousness and death.
2. Hypoxic effects upon the heart and blood vessels may lead to circulatory insufficiency and collapse.
3. Poor oxygen supply to somatic nerve and muscle systems may lead to weakness in otherwise alert victims.

Of these, cerebral hypoxia is by far the most serious incapacitating effect of "hypoxic" gas intoxication, although effects upon heart muscle can be serious in victims with already defective cardiac function [13]. Effects upon somatic nerve and muscle are probably less important. Although the hypoxic atmospheres used in the present study did provide some evidence of poor muscle reflexes and nerve conduction delays, it is unlikely that these were the most serious incapacitating factors, particularly as voluntary muscle can function fairly well under semi-anaerobic conditions for some minutes.

Cerebral Hypoxia

Cerebral tissue is extremely sensitive to interruptions in oxygen supply; failure to supply oxygen at an adequate rate can cause immediate loss of consciousness and irreversible brain damage within seconds. A simple but dramatic example of this is the dizziness and occasional fainting caused by postural hypotension, which occurs when someone stands up rapidly from a sitting or prone position. This results in a pooling of blood for a few seconds in the lower part of the body, causing a reduced cerebral blood flow and thus a reduced oxygen supply to cerebral tissue.

Poor oxygen supply to cerebral nerve cells can result from a number of factors, all of which may be present in the situations under examination. The main considerations are: the cerebral blood supply, the oxygen saturation of arterial blood, the ease with which oxygen is unloaded from the blood into the tissues and the degree of interference with tissue respiration caused by effects upon the cytochrome system.

The interaction between these factors is not always a simple one, but the normal physiological response to hypoxic situations is to maximise the supply of oxygen to cerebral tissue. When an animal is placed in a hypoxic situation as in the case of 10% O_2 exposure, there is a reflex increase in cerebral blood flow caused by cerebral artery dilation. Also, the unloading of oxygen from the blood oxyhaemoglobin is more efficient at a lower arterial and venous PO_2. These factors compensate to a large degree for the decrease in the PO_2 of the inspired air [14]. However, the degree of dilation of the cerebral arteries also depends heavily upon the PCO_2 of the arterial blood. If the PCO_2 falls due to hyperventilation or some other factor, the cerebral arteries constrict and this can result in cerebral hypoxia. This effect has been demonstrated by Meyer and Gotoh [15] who found that mechanical hyperventilation, although supplying excess oxygen, in fact resulted in brain hypoxia in cats and monkeys.

HCN

When HCN is the toxic agent the primary cause of tissue "hypoxia" is the inability of the cells to utilise oxygen due to cytochrome poisoning. However, it has been reported [16] that the hyperventilation caused by HCN inhalation can result in a fall in arterial PCO_2 and cerebral arterioconstriction, which exacerbates the situation.

CO

Carbon monoxide produces hypoxia primarily by combining with haemoglobin and reducing the amount available for the transport of oxygen in the blood. It also reduces the efficiency of unloading of oxygen from oxyhaemoglobin, and may have some effects as a cytochrome poison [17]. A form of interaction between CO and arterial PCO_2 similar to that described for HCN may have implications with regard to CO exposures [18]. Twelve subjects with COHb levels of 25-30% exhibited only minor symptoms, but when they were asked to hyperventilate in air they fell unconscious. Another subject was able to function fairly normally at the exceptionally high level of 55% COHb while resting but collapsed and fell unconscious immediately when he tried to get up to walk. A similar effect is often reported from fire situations when a victim is discovered with high COHb levels in the middle of the bedroom. The victim has apparently awakened in bed and tried to rise and cross the room only to collapse a few feet from the bed (Woolley, personal communication).

Both of these effects are likely to result from a combination of poor cerebral oxygen delivery by CO saturated blood and reductions in cerebral blood flow due in the one case to arterio-constriction following hyperventilation, and in the other to reduced blood pressure caused by postural changes.

CONCLUSIONS

1. It is concluded that the experimental model is suitable for studying the incapacitative effects of fire gases and that the physiological tests used are capable of detecting effects down to near threshold levels.
2. It is suggested that the effects of direct hypoxia in fires are likely to be minimal down to 15% oxygen, but as the levels approach 10% oxygen and below incapacitation is likely within a few minutes.
3. Carbon dioxide levels up to 5% are likely to have little effect, but between 5-10% incapacitation will be increasingly severe.
4. The effects of CO are insidious and there is little incapacitation until the crisis threshold dose is reached, the threshold level depending to some extent upon the posture and activity of the victim.
5. HCN may be important in some fires in producing a rapid "knock-down" causing the victim to remain in the fire and die from CO or other factors. At concentrations below 60 ppm the effects should be minimal.
6. Considerable interactions are likely between fire gases when they are present in combination. The effects of such combinations on time to incapacitation is an important area for further study.

ACKNOWLEDGEMENTS

The author wishes to thank Dr. W. D. Woolley of the Fire Research Station, Borehamwood, U.K., for many helpful discussions during the course of the work.

REFERENCES

1. Bowes, P. C., "Smoke and Toxicity Hazards of Plastics in Fires," *Ann. Occup. Hyg.*, Vol. 17, p. 143 (1974).
2. *U.K. Fire Statistics 1980*, London, Home Office, published annually.
3. Purser, D. A., and Woolley, W. D., "Biological Studies of Combustion Atmospheres," Conference on Smoke and Toxic Gases from Burning

Polymers. London, 6-7 January 1982. QMC Industrial Research Ltd., London. *J. Fire Science*, Vol. 1, p. 118 (1983).

4. Purser, D. A., and Grimshaw, P., "The Incapacitative Effects of Exposure to the Thermal Decomposition Products of Polyurethane Foams," Interflam '82 Conference 30 March-1 April 1982. *Fire Mater*, in press.
5. Cooper, J. R., Blood, F. E., and Roth, R. H., *The Biochemical Basis of Neuroparmacology*, Oxford University Press Inc. (1970).
6. Petajan, J. H., and Grunnet, M. L., "Sequelae of Carbon Monoxide Induced Hypoxia in the Rat," *Arch. Neurol.*, Vol. 33/3, p. 15 (1976).
7. Purser, D. A., "Behaviour of Monkeys at High Levels of Carbon Monoxide Related to Effects in Man," in press.
8. Purser, D. A., Grimshaw, P., and Berrill, K. R., "Intoxication by Cyanide in Fires: A Study in Monkeys Using Polyacrylonitrile," *Arch. Environ Hlth.* in press.
9. Vogel, A. I., "Determination of Cyanides," In: *A Text-Book of Quantitative Inorganic Analysis*, Longmans, Green and Co., p. 327 (1939).
10. Commins, B. T., and Lawther, P. J., "A Sensitive Method for the Determination of Carboxyhaemoglobin in a Finger Prick Sample of Blood," *Brit. J. Industr. Med.*, Vol. 22, p. 139 (1965).
11. Purser, D. A., and Berrill, K. R., "Effects of Carbon Monoxide on Behaviour in Monkeys in Relation to Human Fire Hazard," *Arch. Environ. Hlth.*, Vol. 38, p. 308 (1983).
12. Documentation of the Threshold Limit Values for Substances in Workroom Air. American Conference of Governmental Industrial Hygienists (1980).
13. Kimmerle, G., "Aspects and Methodology for the Evaluation of Toxicological Parameters During Fire Exposure," *JFF/Combustion Toxicology*, Vol. 1, p. 4 (1974).
14. Luft, U. C., *Clinical Cardiopulmonary Physiology*, Grune and Stratton, p. 636 (1969).
15. Meyer, J. S., and Gotoh, F., "Metabolic and Electroencephalographic Effects of Hyperventilation," *Arch. Neurol.*, Vol. 3, p. 539 (1960).
16. Brierley, J. B., Brown, A. W., and Calverley, J., "Cyanide Intoxication in the Rat: Physiological and Neuropathological Aspects," *J. Neurol. Neurosurg. Psychiat.*, Vol. 39, p. 129 (1976).
17. Stewart, R. D., "The Effects of Carbon Monoxide on Man," *JFF/Combustion Toxicology*, Vol. 1, p. 167 (1974).
18. Von Leggenhager, K., "New Data on the Mechanisms of Carbon Monoxide Poisoning," *Acta. Med. Scand.*, Vol. 196, Suppl. 563 (1974).

—————————— 9 ——————————

Acute Inhalation Toxicity of the Smoke Produced by Five Halogenated Polymers

HAROLD L. KAPLAN, ARTHUR F. GRAND and WALTER G. SWITZER

Department of Fire Technology
Southwest Research Institute
San Antonio, Texas

SHAYNE C. GAD

Allied Corporation
Morristown, New Jersey

ABSTRACT

The acute inhalation toxicity of smoke produced by five halogenated polymers used as electrical wire coatings was investigated in this study. The polymers included two chlorofluoropolymers (Halar 500® and Halar 555®) and three fluoropolymers (Teflon 100®, Tefzel 200® and Kynar®). The toxicity of each material was evaluated under flaming and nonflaming combustion using the NBS developmental protocol supplemented with measurement of incapacitation and analyses of the combustion atmospheres for HF, HCl and COF_2.

The smoke produced by Halar 500®, Halar 555® and Kynar® exhibited comparable toxicities in both combustion modes, with LC_{50} values that ranged from 15.1 to 33.3 mg/L. Nonflaming combustion produced high levels of CO and was associated with high incidences of incapacitation and lethality during exposure. Under flaming combustion, small quantities of CO were generated and low incidences of incapacitation and exposure lethality were seen. Lethalities primarily occurred post exposure and were possibly related to the HF and/or HCl evolved. With Teflon 100®, LC_{50} values in flaming (0.08 mg/L) and nonflaming (< 0.05 mg/L) modes were comparable, but toxic effects differed considerably. With Tefzel 200®, differences between combustion modes were observed both in toxicity (LC_{50}'s: flaming 30.2, nonflaming 3.3 mg/L) and in toxic effects. With both Teflon 100® and Tefzel 200®, levels of CO, HF, HCl and COF_2 could not account for the observed effects. The results demonstrate that combustion atmospheres generated by decomposition of halogenated materials at different temperatures may differ considerably in toxicological effects.

136

INTRODUCTION

RECENT FIRES IN public buildings, such as the MGM Grand Hotel, Stouffer's Inn and the Westchase Hilton have increased public awareness of fires and particularly the hazards of smoke inhalation. Although any of a number of factors may cause death, it is now recognized that the primary cause of death in most fires is the inhalation of toxic gases [1].

When combusted, all materials generate toxic gases, the most prevalent of which is carbon monoxide. Other gases may also be produced in toxicologically significant quantities, depending on the chemical structure of the material and on the combustion conditions. An important group of synthetic materials is the halogenated plastics which have found widespread use in construction and furnishings. These materials are of particular toxicological significance because of their capability to generate halogen acids and other respiratory irritants. Only limited data are available on the toxicity of the combustion atmospheres of these materials, except for polytetrafluoroethylene and polyvinyl chloride.

The purpose of this study was to evaluate the acute inhalation toxicity of the smoke produced by five halogenated polymers that are widely used as electrical wire coatings. A secondary objective was to correlate the toxicity with detailed analyses of the smoke in an effort to determine the chemical species responsible for observed toxic effects.

METHODS

Test Materials

The five test materials consisted of the following: two chlorofluoropolymers, Halar 500® (ethylene-chlorotrifluoroethylene) and Halar 555® (blown ethylene-chlorotrifluoroethylene) and three fluoropolymers, Teflon 100® (fluorinated ethylene/fluorinated propylene), Tefzel 200® (ethylene-tetrafluoroethylene) and Kynar® (polyvinylidene fluoride). The halogen content of these materials are: Halar 500® and Halar 555®, 39.4% fluorine and 24.6% chlorine; Teflon 100®, 76.0% fluorine; Tefzel 200®, 59.4% fluorine; Kynar®, 59.4% fluorine.

Animals

Adult male Sprague-Dawley rats (Timco Breeding Laboratories) weighing between 250 and 350 grams were used. Upon receipt, the animals were individually identified by ear tattoo and quarantined for a minimum of two weeks. Animals were allowed free access to water and standard formula Purina rodent chow, except during the exposure.

Combustion

The combustion apparatus was a conductive type furnace which has been described by Potts and Lederer [2]. Nonflaming combustion experiments were conducted at 25 °C below the predetermined autoignition temperature; flaming combustion experiments were conducted at 25 °C above this temperature, except when higher temperatures were necessary to maintain flaming combustion. Additional nonflaming experiments were conducted at 440 °C if the autoignition temperature of a material was greater than 490 °C. The autoignition temperature of each test material was established in a series of experimental trials without animals, by decreasing or increasing the furnace temperature from an initial estimated starting temperature in a mathematical sequence, to an upper limit of 800 °C.

Animal Exposures

Animal exposures were conducted in a 200-L (nominal volume) polymethyl methacrylate chamber in accordance with the U.S. National Bureau of Standards (NBS) developmental test protocol [3], supplemented with measurements of incapacitation and with additional chemical analyses. Prior to the start of each experiment, the animals were weighed and placed in tubular aluminum restrainers which were inserted into the chamber to permit head-only exposure. The experiments were initiated with the addition of a weighed sample of test material into the tared quartz beaker after equilibration of the selected furnace temperature. In each experiment, six animals were exposed in a head-only mode to the combustion atmosphere for a total of 30 minutes. At least five experiments were conducted with varying quantities of each of the five materials in both the flaming and nonflaming combustion modes. In addition, nonflaming combustion experiments at 440 °C included five experiments with each of the two Halar® materials and one experiment each with Teflon 100® , Tefzel 200® and Kynar® . In each experiment, the sample weight was related to the chamber volume and expressed as the sample "charge" or sample concentration in mg/L units.

Toxicity Measurements

Incapacitation. Five of the animals were used for determination of incapacitation by the leg-flexion avoidance response [4] and of lethality. For incapacitation measurement, the rat was positioned to receive an electrical shock (4.6 mA) through its leg upon contact with a metal platform connected to a constant current shocking system. The animal rapidly learns to avoid shock by flexion of its leg and a satisfactory baseline performance level is readily attained. An animal was considered incapacitated when it no longer effectually responded to the shock by flexion of its leg. Each animal's responses and time of responses were recorded on a strip chart.

Lethality. Five test animals were observed for lethality during the exposure and for 14 days post exposure. Animals that expired during exposure or the first hour post exposure were considered exposure lethalities; those that expired after the first hour post exposure were considered post-exposure lethalities. Percent lethality was graphed as a function of sample concentration and the LC_{50} value with confidence limits (30-minute exposure and 14-day post exposure) was derived by the probit method of Finney [5].

Blood Carboxyhemoglobin (COHb) Analysis. The sixth animal was used to obtain a measure of the percent COHb saturation at the median time of incapacitation. Immediately after incapacitation of the third of the five test animals, the sixth animal was removed and a blood sample was obtained by intraorbital venous puncture. In the event none of the animals were incapacitated, a blood sample was obtained from the sixth animal for determination of COHb at the end of the test. Blood samples were also obtained by cardiac puncture from all animals that expired during exposure. Percent COHb saturation was determined by means of an Instrumentation Laboratories Model 282 CO-Oximeter.

Toxic Signs. Animals that survived the exposure were observed for toxic signs such as respiratory abnormalities, ataxia and loss of righting reflex during the first hour post exposure and periodically for 14 days. Animal body weights were recorded pre-exposure and on days 1, 7 and 14 post exposure for comparison with body weight growth patterns of control animals. At the end of 14 days all survivors were necropsied and major organs (heart, lungs, kidneys, liver and spleen) were examined for macroscopic abnormalities. Wet lung weights were obtained for calculation of lung-to-body weight ratios. Mean body and mean lung-to-body weight ratios were statistically analyzed using the Mann-Whitney U test [6].

Combustion Atmosphere Analyses

Analyses of the combustion atmosphere were made continuously for O_2, CO_2 and CO with a closed loop sampling system at a sampling rate of 500 cc/min and the following instrumentation: Beckman OM-11 Analyzer (O_2); Beckman 865 Infrared Analyzers (CO and CO_2).

Combustion atmospheres of all materials were analyzed periodically for hydrogen chloride (HCl) as well as for hydrogen fluoride (HF). The chemical structures of Teflon 100®, Tefzel 200® and Kynar® do not contain chlorine. However, chlorine could be present as a contaminant from the use of chlorine-containing compounds as intermediates in the manufacture of these materials.

Dry sorption tubes were used to sample HCl and HF from the animal exposure apparatus. Approximately 0.5 g of 20-30 mesh soda lime (a mixture of sodium hydroxide and calcium oxide) in a glass tube was used for collection of HCl. Between 0.2 and 0.3 g of 4-8 mesh activated charcoal in a polyolefin tube was used for HF. The tubes were 1/8-inch I.D. and 2-3 inches long. The ends of the tube were plugged with cotton. A suction pump, needle valve and rotameter were employed to pull a specific volume of the combustion atmosphere (in the range 300-900 cc/min) through the sorption tubes. Sufficient samples were obtained over fixed time intervals to enable determination of the maximum concentration and concentration-time product (Ct) for HF and HCl for any given experiment. Due to difficulty in maintaining a specified flow rate, the calibrated rotameter was read frequently over the sample time interval and the total volume of gas was then calculated.

Analysis of HCl was performed by desorbing the soda lime, including the cotton wadding, in 15 mL deionized water. After at least ten minutes, a five mL aliquot was extracted. This sample was adjusted to pH 8-10 (phenolphthalein end point) with 0.1 N HNO_3. Titration of Cl^- was performed to a visual end point (s-diphenyl carbazone) with mercuric nitrate. The titrant was 0.014-0.0014 N, depending on the expected concentration of chloride. Concentration of HCl in the gas phase was calculated from the chloride concentration, knowing the total volume of the atmosphere sampled and the volume of the aqueous solution.

For HF determinations, the charcoal adsorbent, including the cotton wadding, was desorbed in 10 mL of a 1:1 solution of deionized water and Orion Research TISAB ("total ionic strength activity buffer" for fluoride). If the sample concentration fell within the calibrated range of the meter, then the solution was analyzed without dilution. Otherwise, an aliquot was extracted for dilution and this was taken into account in the calculation. A Model 407A (Orion Research) ion electrode meter and a 94-09 Fluoride Ion Electrode with a 90-02 Reference Electrode were employed to measure the concentration of fluoride in the solution. The meter was calibrated in the range 0.5-2.0 mg F^-/L solution. Calculations were performed as for HCl, using the concentration of F^- and the volumes of the atmosphere sampled and of the solution analyzed. Combustion atmospheres were also sampled in selected experiments by the use of evacuated glass sampling vessels and analyzed for carbonyl fluoride (COF_2) by gas chromatography.

RESULTS AND DISCUSSION

The results of relevant experiments with each of the five materials under flaming, nonflaming and 440 °C (nonflaming) combustion conditions are summarized in Tables 1 to 5. Gas evolution curves for CO, HF and HCl for some of these experiments are shown in Figures 1 to 6.

Halar 500®

An LC_{50} value of 15.1 mg/L for flaming combustion of Halar 500® was estimated from the three experiments shown in Table 1. The results of three additional experiments were anomalous due to difficulties encountered in sustaining flaming combustion of the material. Inclusion of these results would have increased the LC_{50} value slightly. In the three experiments in which flaming combustion was sustained, only one animal was incapacitated and there were no exposure lethalities. In each of two experiments, three animals died post exposure. The low incidences of incapacitation and lethality during exposure are consistent with the analytical data for CO and COHb in these experiments. Maximum CO concentrations ranged from 490 to 1030 ppm and Ct products for CO were from 13,710 to 29,190 ppm-min. Blood COHb saturations were less than 30 percent in exposed animals at the end of the tests. These concentrations, Ct products and COHb levels are considerably below threshold values for incapacitation and exposure lethality for pure CO, as determined by this laboratory. In pure gas studies with CO, Ct products of approximately 35,000 and 120,000 ppm-min. and COHb levels of approximately 60 and 70% were necessary for incapacitation and lethality, respectively [7]. Significant quantities of HF (750–1570 ppm maximum) and HCl (1160–3010 ppm maximum) were generated by this material. In pure gas studies, investigators have determined 5-min LC_{50} values of 18,200 ppm (HF) and 40,989 ppm (HCl) and 60-min LC_{50} values of 1395 ppm (HF) and 3124 ppm (HCl) [8,9]. These concentrations and exposure times are equivalent to Ct products of approximately 80,000 and 90,000 ppm-min (HF) and 185,000 and 205,000 ppm-min. (HCl). Comparable Ct products have been obtained from LC_{50} values determined in studies of HCl in this laboratory, with some post-exposure deaths resulting from doses with Ct products as low as 54,000 ppm-min [7]. Therefore, in view of the fact that the toxicities of HF and HCl are additive [9], the combined Ct products for the two gases in some of the Halar 500® experiments were sufficiently large to account for the post-exposure deaths.

The LC_{50} value for nonflaming combustion was determined to be 20.1 (18.4–22.0) mg/L. Nonflaming combustion atmospheres of Halar 500® yielded more consistent results than those of the flaming mode. Predominant toxic effects of these atmospheres were incapacitation and exposure lethality. These effects are consistent with the CO and COHb analytical data. Maximum CO concentrations and Ct products for CO ranged from 2670 to greater than 5000 ppm and from 71,330 to greater than 132,810 ppm-min, respectively. Blood COHb levels

Table 1. Summary of animal exposure and analytical data
Halar 500—autoignition temperature 655°C.

| Sample Conc., mg/L | Percent Weight Loss | No. of Animals Incapacitated | % Lethality | | % COHb | | | Carbon Monoxide | | Hydrogen Fluoride | | Hydrogen Chloride | | COF₃ |
			Exposure	Total	Incapacitation	End of Test	Mean Expo. Deaths	Max., ppm	Ct ppm-min	Max., ppm	Ct ppm-min	Max., ppm	Ct ppm-min	Max., ppm
Flaming, 680-750°C — LC₅₀ approximately 15.1 mg/L														
7.5	100	0	0	0	—	19.8	—	490	13710	750	7450	1160	16980	—
15	100	1	0	60	—	29.3	—	1030	29190	1540	12450	2510	37450	—
17.5	100	0	0	60	—	29.1	—	890	24290	1570	16870	3010	47180	—
Nonflaming, 630°C — LC₅₀ 20.1 (18.4-22.0) mg/L														
15	100	1	0	0	—	66.5	—	2670	71330	450	3250	2090	25370	—
17.5	100	2	20	20	—	74.3	77.0	3290	81870	840	9100	2100	29220	—
20	100	5	20	60	65.2	—	79.3	4010	101170	810	8110	2740	33430	—
22.5	100	5	60	60	67.0	—	78.7	4500	112820	1070	8750	2630	39970	—
25	100	5	100	100	71.3	—	81.6	(5000)	(132810)	1125	11780	3360	54200	ND
440°C — LC₅₀ 27.8 (19.3-38.9) mg/L														
17.5	<90.7	0	0	0	—	53.5	—	2500	40970	1920	22910	2320	41830	—
20	81.5	0	0	0	—	20.4	—	2670	26320	1140	18500	1370	24660	—
22.5	<99	0	0	60	—	57.6	—	3110	48960	3690	30680	2910	43270	—
27.5	90	4	20	60	76.1	—	—	4270	88670	2940	32800	4030	72100	410
30	89	2	0	40	—	69.7	—	4290	58980	3100	28440	4020	56324	—

ND = not detectable.

142

Table 2. Summary of animal exposure and analytical data
Halar 555—autoignition temperature 655°C.

Sample Conc., mg/L	Percent Weight Loss	No. of Animals Incapacitated	% Lethality Exposure	Total	Incapacitation	% COHb End of Test	Mean Expo. Deaths	Carbon Monoxide Max., ppm	Ct ppm-min	Hydrogen Fluoride Max., ppm	Ct ppm-min	Hydrogen Chloride Max., ppm	Ct ppm-min	COF₂ Max., ppm
Flaming, 800°C—LC₅₀ approximately 20 mg/L														
14	100	0	0	0	—	15.7	—	490	12140	720	6690	2220	22900	ND
17.5	100	0	0	0	—	27.9	—	750	18840	1180	9160	2710	32520	—
20	100	0	0	20	—	27.8	—	970	22860	1660	8330	3250	41600	—
20	100	0	0	100	—	30.5	—	1000	24270	1180	10050	3100	33210	—
20	100	0	0	40	—	12.4	—	810	19700	2100	11850	2970	33720	—
Nonflaming, 630°C—LC₅₀ 28.9 (20.3-41.1) mg/L														
22.5	100	3	40	40	—	76.7	—	3530	91430	1180	10010	2880	30730	—
27.5	100	2	0	0	—	80.6	—	3860	97820	1290	10710	3320	45400	40
27.5	100	4	20	40	67.9	—	78.2	3920	100140	2930	22060	3180	47960	—
27.5	100	4	40	60	77.7	—	78.6	3830	97940	1280	8880	3240	40970	—
32	100	4	40	60	82.2	—	82.9	4680	116800	2200	14620	3230	46720	—
440°C—LC₅₀ 33.3 (31.3-35.1) mg/L														
22.5	86.7	0	0	0	—	62.2	—	2450	41810	1770	16900	3150	51160	—
30	86.0	0	20	20	—	74.6	—	3170	59230	2620	23090	4120	62040	140
32.5	87.8	0	20	20	—	74.2	—	3730	68750	2550	20470	4450	63850	—
32.5	86.8	2	40	40	—	76.5	—	3470	62910	3080	25730	3260	39010	—
35	88.2	1	80	80	—	81.4	—	4010	77610	1940	12780	4780	65220	—

ND = not detectable.

Table 3. Summary of animal exposure and analytical data
Teflon 100—autoignition temperature 565°C.

Sample Conc., mg/L	Percent Weight Loss	No. of Animals Incapacitated	% Lethality Exposure	% Lethality Total	% COHb Incapacitation	% COHb End of Test	% COHb Mean Expo. Deaths	Carbon Monoxide Max., ppm	Carbon Monoxide Ct ppm-min	Hydrogen Fluoride Max., ppm	Hydrogen Fluoride Ct ppm-min	Hydrogen Chloride Max., ppm	Hydrogen Chloride Ct ppm-min	COF$_3$ Max., ppm
Flaming, 590°C—LC$_{50}$ 0.075 (0.03–0.27) mg/L														
0.032	100	0	0	20	—	1.5	—	ND	—	10	260	<10	120	—
0.04	100	0	0	0	—	0.7	—	ND	—	20	240	<10	60	—
0.3	100	0	0	100	—	1.5	—	50	—	110	1670	ND	—	—
1.0	100	4	100	100	1.1	—	1.4	50	1310	330	4210	NA	—	70
10.0	100	5	100	100	6.3	—	5.6	440	12660	1060	15520	100	3000	—
Nonflaming, 540°C—LC$_{50}$ <0.05 mg/L														
0.011	100	0	0	100	—	1.5	—	ND	—	<10	70	ND	—	—
0.02	100	0	0	20	—	1.4	—	ND	—	10	170	ND	—	—
0.055	100	0	0	100	—	1.6	—	ND	—	10	190	ND	—	—
0.2	100	0	0	60	—	1.4	—	ND	—	40	920	ND	—	—
0.5	100	1	0	60	—	2.1	—	ND	—	70	1590	20	660	—
1.0	100	0	0	100	—	1.0	—	ND	—	160	3780	10	360	150
440°C														
1.0	13.9	0	0	20	—	1.5	—	—	—	20	360	ND	—	—

ND = not detectable.
NA = not analyzed.

144

Table 4. Summary of animal exposure and analytical data
Tefzel 200—autoignition temperature 555°C.

Sample Conc., mg/L	Percent Weight Loss	No. of Animals Incapacitated	% Lethality Exposure	Total	Incapacitation	% COHb End of Test	Mean Expo. Deaths	Carbon Monoxide Max., ppm	Ct ppm-min	Hydrogen Fluoride Max., ppm	Ct ppm-min	Hydrogen Chloride Max., ppm	Ct ppm-min	COF₂ Max., ppm
Flaming, 590°C — LC₅₀ 30.2 (22.8-40.0) mg/L														
3.0	100	0	0	0	—	4.0	—	280	7780	490	7190	<10	60	—
10.0	100	0	0	0	—	10.0	—	260	7396	420	3760	<10	90	—
20.0	100	0	20	20	—	16.3	18.8	380	10970	670	6800	10	270	60
30.0	100	0	40	40	—	39.7	38.0	990	28520	770	8540	20	280	—
40.0	100	3	40	80	24.6	—	34.7	860	19450	940	9550	50	550	—
Nonflaming, 530°C - LC₅₀ 3.33 (2.9-3.8) mg/L														
1.0	100	0	0	0	—	3.5	—	50	1210	90	1570	<10	60	—
2.5	100	1	0	0	—	4.2	—	120	3330	210	3000	ND	—	—
3.0	100	1	0	40	—	5.0	—	100	2790	280	4120	ND	—	—
4.0	100	2	40	80	—	7.3	6.9	200	5420	380	5210	<10	120	—
7.0	100	4	100	100	8.8	—	7.7	310	8480	590	9270	<10	220	650
440°C														
3.3	99	0	0	0	—	5.2	—	90	2330	90	1910	ND	—	—

ND = not detectable.

145

Table 5. Summary of animal exposure and analytical data
Kyner—autoignition temperature 630°C.

Sample Conc., mg/L	Percent Weight Loss	No. of Animals Incapacitated	% Lethality Expo-sure	% Lethality Total	Incapaci-tation	% COHb End of Test	Mean Expo. Deaths	Carbon Monoxide Max., ppm	Carbon Monoxide Ct ppm-min	Hydrogen Fluoride Max., ppm	Hydrogen Fluoride Ct ppm-min	Hydrogen Chloride Max., ppm	Hydrogen Chloride Ct ppm-min	COF_2 Max., ppm
Flaming, 655–750°C – LC_{50} 27.3 (17.9–41.7) mg/L														
15	100	2	0	20	—	22.2	—	500	13030	1790	14570	50	940	—
20	100	3	0	20	45.6	—	—	1890	45730	1760	13040	280	1610	—
30	100	0	0	20	—	39.5	—	1480	38120	4160	37390	30	470	—
35	100	0	0	100	—	36.7	—	1320	30060	3910	36920	30	750	—
40	100	0	0	80	—	ND	—	1540	36010	4390	30550	40	830	860
Nonflaming, 605°C – LC_{50} 24.3 (19.1–31.2) mg/L														
12.5	100	0	0	0	—	65.3	—	2470	64380	910	8060	90	—	—
20	100	5	0	0	54.3	—	—	4220	105410	1700	17580	90	1090	—
23	100	5	20	20	72.9	—	79.4	4620	119750	2540	18590	30	670	—
24.5	100	5	40	40	70	—	82.8	(5000)	(1132030)	3060	20300	40	760	—
26	100	5	100	100	63.1	—	83.4	(5000)	(1131300)	3440	28940	60	1170	200
440°C														
24.3	82	0	0	0	—	50.8	—	3610	43410	3580	42080	40	970	—

146

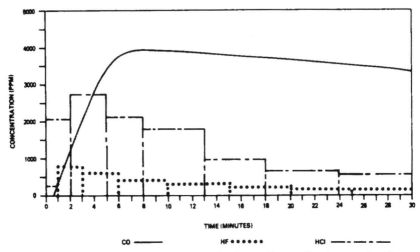

Figure 1. Gas evolution from Halar 500®, 20 mg/L, nonflaming combustion.

were 65.2 to 71.3 percent at incapacitation and 77.0 to 81.6 percent at death. Maximum concentrations of HF and HCl and Ct products were generally lower in nonflaming than flaming combustion atmospheres, even though sample concentrations were greater in some of the nonflaming tests.

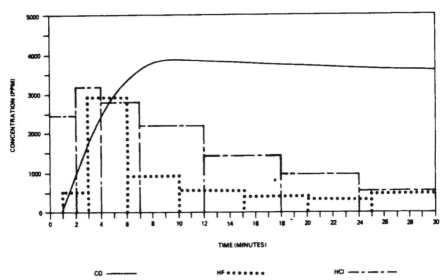

Figure 2. Gas evolution from Halar 555®, 27.5 mg/L, nonflaming combustion.

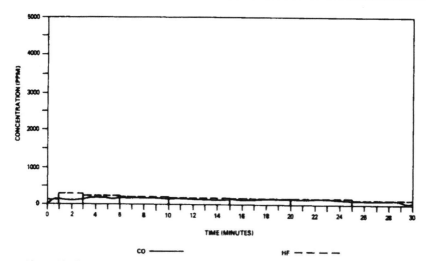

Figure 3. Gas evolution from Tefzel 200®, 3 mg/L, nonflaming combustion.

The LC$_{50}$ value for Halar 500® under nonflaming combustion at 440 °C was determined to be 27.8 (19.3–38.9) mg/L. Considerable quantities of CO were generated in this mode but the evolution was much slower than under nonflaming combustion at 630 °C. Consequently, maximum CO concentrations were high but these concentrations were

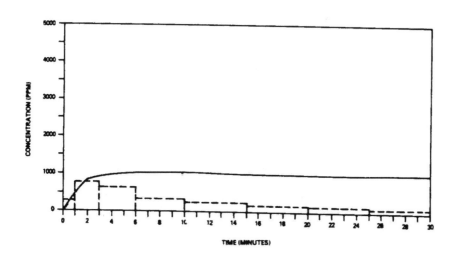

Figure 4. Gas evolution from Tefzel 200®, 30 mg/L, flaming combustion.

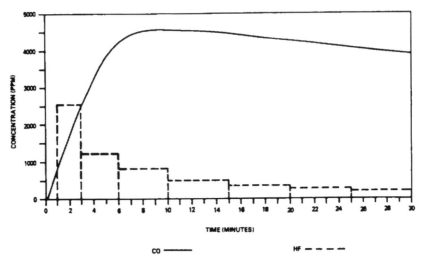

Figure 5. Gas evolution from Kynar® , 23 mg/L, nonflaming combustion.

not attained until the end of the test, so that Ct products were generally less than one-half the values of nonflaming (630 °C) combustion atmospheres. The analytical results are consistent with the lower incidences of incapacitation and exposure lethality and with the lower blood COHb levels at the end of the tests. Concentrations of HF and

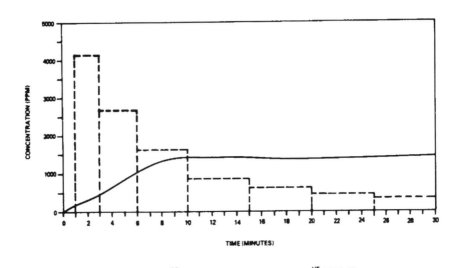

Figure 6. Gas evolution from Kynar® , 30 mg/L, flaming combustion.

HCl were generally greater in this mode than either in the nonflaming (630 °C) or flaming modes and combined Ct products for the two gases in some experiments were sufficiently large to account for the post-exposure deaths.

In general, animals exposed to the atmospheres produced by each of the three combustion modes exhibited toxic signs such as corneal opacities and respiratory abnormalities (dyspnea, gasping) characteristic of exposure to highly irritant and corrosive gases. Depressed body weights were evident at 7 and 14-days post exposure in almost all experiments in each combustion mode. Mean lung/body weight ratios were significantly higher ($p < 0.05$) than control values in some of the experiments in each mode but a dose-effect relationship was not evident. There was no evidence of other macroscopic abnormalities in major organs of surviving animals at the end of the 14-day post-exposure period.

Halar 555®

Five experiments under flaming combustion of Halar 555® at sample concentrations of from 14 to 20 mg/L resulted in lethalities of from 0 to 100 percent. However, difficulties were encountered in sustaining flaming combustion of the material in some of the experiments. Because of the variability in combustion conditions, the toxicological and analytical results did not show a well-defined relationship to sample concentration and it was not possible to derive an exact LC_{50} value. An LC_{50} value of approximately 20 mg/L for flaming combustion may be estimated from the data. Flaming combustion atmospheres did not incapacitate any animals and all lethalities occurred post exposure. The absence of incapacitation and of exposure lethalities is consistent with the analytical data for CO and COHb. Maximum CO concentrations in these experiments did not exceed 1000 ppm, the maximum Ct product was 24,260 ppm-min, and the maximum COHb blood saturation at the end of the tests was only 30.5 percent. Significant quantities of HF (720–2100 ppm maximum) and of HCl (2220–3250 ppm maximum) were evolved by this material and the combined Ct products for the two gases in some experiments were consistent with the occurrence of post-exposure lethalities.

The LC_{50} value for nonflaming combustion of Halar 555® was determined to be 28.9 (20.3–41.1) mg/L. The predominant toxic effects of this combustion mode were incapacitation of animals and both exposure and post-exposure lethalities. In contrast to the flaming mode, large quantities of CO were evolved by nonflaming combustion of Halar 555®. Maximum CO concentrations, Ct products and COHb blood levels were sufficiently high to account for the toxic effects observed during exposure. Maximum concentrations and Ct products for both HF and HCl were generally larger in nonflaming than flaming combustion experiments, probably because larger sample quantities were combusted in the former mode.

The LC_{50} value of nonflaming combustion at 440 °C was determined to be 33.3 (31.3–35.1) mg/L. At 440 °C, decomposition of the material was less complete than under nonflaming at 630 °C and there was a slower evolution of CO. Although maximum CO concentrations were relatively high, Ct products were much lower than the 630 °C values because CO concentrations did not reach maximum values until towards the end of the tests. The CO analytical data are consistent with the low incidence of incapacitation and absence of exposure lethalities. In contrast to CO, maximum concentrations and Ct products of HF and HCl were generally higher in the 440 °C than 630 °C combustion atmospheres. The combined Ct products for the two gases were sufficiently large to account for the high incidence of post-exposure lethalities observed in this mode.

Animals exposed to flaming or nonflaming combustion atmospheres exhibited toxic signs (corneal opacities, respiratory abnormalities) characteristic of exposure to highly irritant and corrosive gases. Animal body weights were significantly ($p < 0.05$) depressed at seven days in almost all experiments and at 14 days in some of the experiments. Mean lung/body weight ratios in exposed animals were significantly greater ($p < 0.05$) than control means in almost all experiments. However, there was no evidence of other macroscopic abnormalities in the major organs of surviving animals at 14-days post exposure.

Teflon 100®

The LC_{50} value for flaming combustion of Teflon 100® was determined to be 0.075 (0.03–0.27) mg/L. At lower sample concentrations, the combustion atmospheres did not incapacitate animals and all lethalities occurred post exposure. At the two highest sample concentrations, almost all animals were incapacitated and there was 100-percent lethality during each exposure. The analytical data from these experiments indicate that the quantities of CO, HF, HCl and COF_2 were insufficient to account for either incapacitation, exposure lethalities or post-exposure lethalities. In none of the tests did the maximum CO concentration exceed 440 ppm or the maximum Ct product exceed 12,660 ppm-min. Blood COHb concentrations at incapacitation, at the end of the tests and in exposure lethalities were equivalent to unexposed animal baseline values. Maximum concentrations for HF and HCl did not exceed 1060 and 100 ppm, respectively, and maximum Ct products for these gases were only 15,520 and 3000 ppm-min, respectively. In the one experiment in which COF_2 was analyzed, the maximum concentration was only 70 ppm.

In the nonflaming combustion mode, six experiments were conducted with a range of sample concentrations of Teflon 100® of from 0.011 to 1.0 mg/L. Some anomalies were observed in lethality results at the lower sample concentrations, probably because of methodology limitations with small sample quantities. Therefore, it was not possible to derive a well-defined dose-response curve and an exact LC_{50} value from the data. However, an LC_{50} value of < 0.05 mg/L was estimated for this material in the nonflaming mode. The nonflaming combustion atmospheres did not incapacitate animals, with one exception, and did not cause any lethalities during exposure. Each sample concentration resulted in at least one death but all lethalities occurred post exposure, almost invariably during the first 24 hours. The absence of toxic effects (incapacitation, lethality) during exposure is consistent with the analytical data for CO and COHb. Carbon monoxide was not detected in any of the nonflaming combustion atmospheres and all blood COHb levels at the end of the tests were in the baseline range. As for the post-exposure lethalities, insufficient quantities of HF, HCl or COF_2 were evolved to account for these effects.

In contrast to the two Halar® materials, neither corneal opacities nor other evidence of exposure to highly irritant and corrosive gases were observed in animals exposed to Teflon 100® combustion atmospheres (flaming or nonflaming). Except for labored or rapid breathing, post-exposure toxic signs were minimal and, except for lethality, were not evident after the first few hours following exposure. Data relevant to body weight, lung/body weight ratios and pathology were limited because of the high incidence of lethalities in these experiments. However, macroscopic abnormalities were not evident in major organs of surviving animals at 14-days post exposure.

Previous studies have shown that the thermal decomposition products of other Teflon® polymers (polytetrafluoroethylene) vary with the temperature as well as the combustion mode [10]. When these materials were pyrolyzed below 500 °C, various products were produced at different temperature ranges and included HF, COF_2, particulate, tetrafluoroethylene (TFE), hexafluoropropylene (HFP), perfluoroisobutylene (PFIB) and hexafluororethane (HFE). Above 550 °C, the entire product is thought to be COF_2. In this study, only small quantities of COF_2 were measured when Teflon 100® was combusted at 590 °C (and at 540 °C); however, this material is a mixed polymer of fluorinated ethylene and fluorinated propylene and is not identical structurally to polytetrafluoroethylene.

Tefzel 200®

The LC_{50} value for flaming combustion of Tefzel 200® was determined to be 30.2 (22.8–40.0) mg/L. Flaming combustion atmospheres did not incapacitate animals, except at the highest sample concentration (40 mg/L). Also, all lethalities occurred during exposure, except for the 40 mg/L experiment in which lethalities occurred both during and after exposure. The analytical data indicate that insufficient quantities of CO, HF, HCl and COF_2 were evolved by the material to account for the observed incapacitation, exposure lethalities or post-exposure lethalities. In addition, the low COHb levels are consistent with the analytical data and are considerably below the levels at which incapacitation and exposure lethality occur.

Nonflaming combustion of Tefzel 200® generated considerably more toxic atmospheres than the flaming mode. The LC_{50} value for nonflaming combustion was determined to be 3.33 (2.9–3.8) mg/L. In these experiments, the incidence of incapacitation was higher than in the flaming combustion experiments and lethalities occurred both during and after exposure. From the analytical data and COHb measurements, it is evident that insufficient quantities of CO, HF, HCl and COF_2 were evolved by the material to account for the observed toxic effects.

In general, toxic signs exhibited by animals after exposure to combustion atmospheres produced by Tefzel 200® were minimal and were not characteristic of exposure to highly irritant and corrosive gases. Animal growth patterns were depressed at 7 and 14-days post exposure in some of the experiments in both the flaming and nonflaming modes. Mean lung/body weight ratios of surviving animals were significantly higher ($p < 0.05$) than control means in all of the flaming and in some of the nonflaming experiments. A dose-related effect was suggested by the data, although the data were limited. There was no other evidence of macroscopic abnormalities in the major organs of surviving animals at 14-days post exposure.

Kynar®

The LC_{50} value for flaming combustion of Kynar® was determined to be 27.3 (17.9–41.7) mg/L. Flaming combustion atmospheres did not generally incapacitate animals, although there were exceptions at the lowest sample concentrations, and did not cause exposure lethalities. Lethalities invariably occurred on the first post-exposure day. The low incidence of incapacitation and absence of exposure lethalities are consistent with the analytical data for CO and COHb. Maximum CO concentrations ranged from 500 to 1890 ppm and Ct products from 13,030 to 45,730 ppm-min. Blood COHb concentrations at the end of the tests did not exceed 39.5 percent. Large quantities of HF and small quantities of HCl were evolved by flaming combustion of Kynar®. In three combustion atmospheres, the maximum HF concentration was greater than 3000 ppm. It is possible that the large quantities of HF evolved by this material contributed to the predominant toxic effects, i.e., the post-exposure lethalities. In the one experiment in which COF_2 was analyzed, a maximum concentration of 860 ppm was measured.

An LC_{50} value of 24.3 (19.1–31.2) mg/L was determined for nonflaming combustion of Kynar®. In contrast to the flaming mode, the predominant toxic effects of nonflaming combustion were incapacitation and exposure lethalities. These toxic effects are consistent with the large quantities of CO evolved by the material. Maximum CO concentrations in most experiments exceeded 4000 ppm and Ct products were greater than 100,000 ppm-min. Blood COHb levels were in the ranges at which incapacitation and exposure lethalities would be anticipated. Large quantities of HF and small amounts of HCl were also generated by nonflaming combustion of Kynar®.

In general, clinical signs (corneal opacities, eye irritation and respiratory distress) exhibited by animals exposed to flaming and nonflaming combustion atmospheres of Kynar® were consistent with exposure to highly irritant and corrosive gases. Depressed growth patterns were evident in animals from most of the flaming and nonflaming combustion experiments at both 7 and 14 days post-exposure. Mean lung/body weight ratios of exposed animals were significantly ($p < 0.05$) higher than control means in three flaming combustion experiments. There was no evidence of other macroscopic abnormalities in the major organs of surviving animals at 14-days post exposure.

CONCLUSIONS

It is well recognized that the combustion products of a material may vary considerably with the temperature at which the material is subjected to thermal decomposition [11]. Because these products may vary, differences in toxic potency and toxicological effects may occur when a material is decomposed at different temperatures. In this study, atmospheres produced by decomposition of each material at different temperatures did not differ greatly in toxic potency, except for Tefzel 200® (with this material, a ten-fold difference in toxic potency was observed between the flaming and nonflaming combustion modes). However, atmospheres produced at the different temperatures did differ in toxicological effects. With three of the materials (Halar 500®, Halar 555® and Kynar®), it was possible to determine the gases responsible for the toxic potencies and toxic effects of the combustion atmospheres. In the case of Teflon 100® and Tefzel 200®, toxic potencies and toxic effects were not correlatable with the results of chemical analyses of the atmospheres.

REFERENCES

1. Crane, C.R., Sanders, D.C., et al., "Inhalation Toxicology: I. Design of a Small-Animal Test System II. Determination of the Relative Toxic Hazards of 75 Aircraft Cabin Materials," Department of Transportation, Federal Aviation Administration, Report No. FAA-AM-77-9 (March 1977).

2. Potts, W.J., and Lederer, T.S., "A Method for Comparative Testing of Smoke Toxicity," *Journal of Combustion Toxicology*, *4*, pp. 114-162 (1977).
3. National Bureau of Standards Developmental Protocol of January 18, 1979, including modifications to the protocol made on August 11, 1980.
4. Packham, S.C., "Behavior and Physiology: Tools for the Assessment of Relative Toxicity," Paper presented at 17th National Cellular Plastics Conference on Safety and Product Liability, Washington, D.C. (November 18-20, 1974).
5. Finney, D.J., *Probit Analysis* (2nd ed.), Cambridge University Press (1964).
6. Siegal, S., *Nonparametric Statistics for the Behavioral Sciences*, McGraw-Hill Book Company (1956).
7. Switzer, W.G., and Kaplan, H.L., Southwest Research Institute, Unpublished Results.
8. DiPasquale, L.C., and Davis, H.V., "The Acute Toxicity of Brief Exposures to Hydrogen Fluoride, Hydrogen Chloride, Nitrogen Dioxide, and Hydrogen Cyanide Singly and in Combination with Carbon Monoxide," *AMRL-TR-17-120*, pp. 279-290 (1971).
9. Wohlslagel, J., DiPasquale, L.C., and Vernot, E.H., "Toxicity of Solid Rocket Motor Exhaust-Effects of HCl, HF and Alumina on Rodents," *AMRL-TR-75-125 No. 14* (1975).
10. Waritz, R.S., and Kwon, B.K., "The Inhalation Toxicity of Pyrolysis Products of Polytetrafluoroethylene Heated Below 500 Degrees Centigrade," *American Industrial Hygiene Association Journal*, Volume 29, pp. 19-26 (January-February, 1968).
11. Kaplan, H.L., Grand, A.F., and Hartzell, G.E., *Combustion Toxicology: Principles and Test Methods*, Technomic Publishing Company, Incorporated, Lancaster, Pennsylvania (1983).

Toxic Hazards of Acrolein and Carbon Monoxide During Combustion

TOKIO MORIKAWA

Fire Research Institute
Mitaka, Tokyo 181
Japan

ABSTRACT

Low molecular weight flammable liquid materials and sucrose, in place of polymeric materials such as polyethylene, polypropylene, cellulose, etc., were combusted in a tube furnace under different burning conditions. Evolution of acrolein and carbon monoxide were determined using gas chromatographic analysis. It was found that acrolein may have a greater lethal role than carbon monoxide under certain burning conditions.

INTRODUCTION

FIRES OFTEN CAUSE death due to gas poisoning even when amounts of combustibles burned in fire are small.

Although a great number of toxicants are possibly evolved in fire, carbon monoxide (CO) is one of the most important toxicants which most likely causes death in fire. However, acrolein is another important toxicant depending on the fuel and the burning condition [1].

In the present study, when low molecular weight flammable liquid materials and sucrose were combusted or pyrolyzed in a tube furnace, the evolution of acrolein and CO was determined relative to the furnace temperature and the air supply rate.

Using the experimental results, the yield of acrolein in relation to the molecular structure of fuels, and a comparison of toxic hazards due to acrolein and CO are described.

EXPERIMENTAL

The apparatus used for combustion and/or pyrolysis is shown in Figure 1. The electric tube furnace was 60 cm long, and the quartz tube

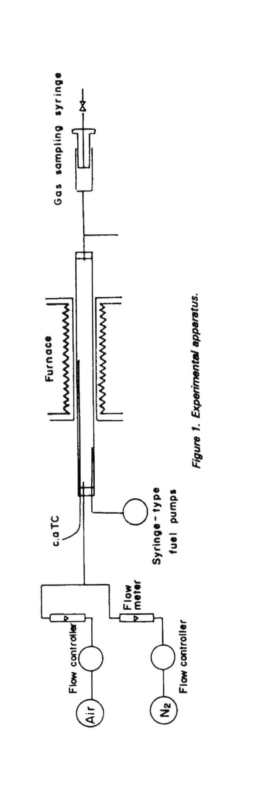

Figure 1. Experimental apparatus.

fitted into the furnace was 3.6 cm in inside diameter. Low molecular weight flammable liquids, n-hexane, iso-hexane, 2,3 dimethyl butane and 2,2 dimethyl butane were used for fuels. These are considered to be structurally close to the decomposition products of e.g. polyethylene and polypropylene. Oxygen-containing compounds, n-propyl alcohol, n-amyl alcohol and a 50% sucrose water solution were also used. These were chosen, because the first is structurally close to acrolein, the second is an oxygen-containing compound which has carbon atoms only one less than hexanes and has a relatively low boiling point, and the last is structurally close to cellulose. Each fuel was combusted in air or nitrogen-diluted air, or pyrolyzed in nitrogen atmosphere. The flow rate of these gases was controlled with a diaphram-type flow controller. The fuel was introduced 5 cm upstream the inlet of the furnace into the quartz tube with a syringe-type pump at a constant rate. The quartz tube at the point of the fuel introduction was heated with a bunzen burner flame to prevent a fuel vapor from condensing on the wall of the quartz tube. The temperature of the furnace at its center was monitored with a chromel-alumel thermocouple and automatically controlled with a temperature controller.

The effluent gas was sampled with a gas sampling syringe. The sampled gas was gas chromatographically analyzed using a Porapak Q-packed column of 3 m at 110 °C for acrolein, and a molecular sieve 5A-packed column of 3 m at 80 °C for inorganic gases. From the determined concentrations, the yields of acrolein and CO per mass of fuel was calculated based on N_2 concentration, on the assumption that N_2 is unchanged during combustion.

RESULTS AND DISCUSSION

Evolution of Acrolein and CO from Combustion of N-hexane and its Isomers

The yields of acrolein and CO (mg/g of sample) from the combustion of n-hexane are shown in Figure 2 (a) and (b) as a function of the air supply rate, with the furnace temperature as a parameter. It can be seen from this figure that the yields of both acrolein and CO were large when there was little or no flaming due to low temperature (350–450 °C) or oxygen deficiency (up to an equivalence ratio of fuel to oxygen, ϕ, of about 0.25). However, when the air supply rate was 0.5 l/min (or ϕ was as low as 0.125), the yield of CO increased and that of acrolein decreased as the furnace temperature was increased to 600 °C or over. The reason for this may be that acrolein is unstable and/or less readily formed, whereas CO is stable at high temperature. At temperatures of 450 °C or below, where no flaming occurred, the yields of acrolein and CO increased with increasing air supply rate, but decreased when the air supply rate reached 2 l/min or higher. This decrease (with increased

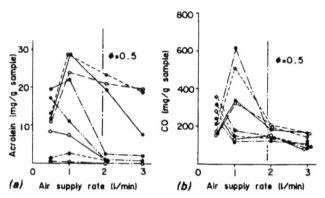

Figure 2. Yields of acrolein and CO from combustion of n-hexane. Furnace temperature: O *350°C,* ● *400°C,* ☉ *450°C,* ◑ *500°C,* ◔ *550°C,* ◇ *600°C,* ◈ *650°C,* ◈ *700°C. Fuel supply rate: 0.3 g/min.* ◆*: Equivalence ratio of fuel to oxygen.*

air supply rate) may have been *caused by the decrease in residence time* of the reactants in the furnace.

Effects of oxygen concentration diluted with nitrogen gas on the yields of acrolein and CO were also determined. Combustion of n-hexane was conducted at different furnace temperatures with the supply rate of air fixed at 1 *l*/min and that of nitrogen gas varied. The yields of acrolein and CO as a function of the nitrogen supply rate is shown in Figures 3 (a) and (b). At 450 °C or below, the yields of both acrolein and CO decreased with increasing nitrogen supply rate. However, at temperatures between 500 and 650 °C, the yield of CO in-

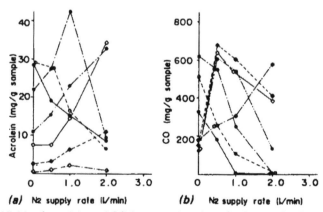

Figure 3. Yields of acrolein and CO from combustion of n-hexane in nitrogen-diluted air atmosphere. Furnace temperature: ● *400°C,* ☉ *450°C,* ◑ *500°C,* ◔ *550°C,* ◇ *600°C,* ◈ *650°C,* ◈ *700°C. Air supply rate: 1.0 l/min. Fuel supply rate: 0.3 g/min.*

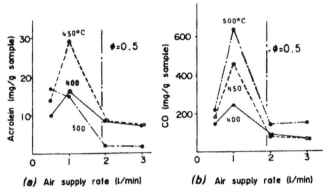

Figure 4. Yields of acrolein and CO from combustion of iso-hexane. Fuel supply rate: 0.3 g/min.

creased to a maximum and then decreased, while that of acrolein (except at 500 °C) increased steadily when the nitrogen supply rate was increased to 2 l/min. The yield of acrolein would be expected to decrease with high nitrogen supply rates. The excessively high nitrogen supply rates apparently lead to a decrease in residence time of reactants in the furnace due to the increased flow speed. At 700 °C, the yield of CO increased with increasing nitrogen supply rate, while that of acrolein was negligibly small at any nitrogen supply rate. The reason for this, as mentioned before, is that in non-flaming combustion CO is stable at high temperature, while acrolein is unstable and/or less easily formed.

The yields of acrolein and CO from combustion of isomers of n-hexane (iso-hexane, 2,3 dimethyl butane and 2,2 dimethyl butane) as a function of the air supply rate are shown in Figure 4 through Figure 6. The maximum yields of both acrolein and CO occurred when ϕ was approximately 0.25, as in the case of n-hexane. The maximum yield of acrolein from 2,2 dimethyl butane (e.g., at 450 °C) was lower than the maximum yields of acrolein from n-hexane, iso-hexane or 2,3 dimethyl butane under similar conditions. On the other hand, the maximum yield of CO was not much different among the four.

In order for acrolein to be evolved as a combustion product, 3-carbon chains must be formed. Normal hexane undergoes random scissions between carbon atoms with oxygen attack. For iso-hexane and 2,3 dimethyl butane, the probability of the β carbon atom being attacked by oxygen is high. Thus, from these three hexanes, straight chain hydrocarbons with 3 carbon atoms can be formed. However, in the case of 2,2 dimethyl butane, the carbon atom at the γ position reacts with oxygen more readily than carbons at other positions; therefore, the chance is small that straight chain hydrocarbons with 3 carbon

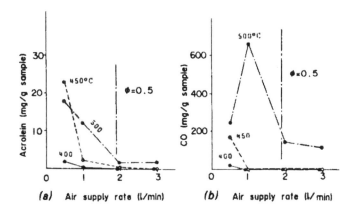

Figure 5. Yields of acrolein and CO from combustion of 2,3 dimethyl butane. Fuel supply rate: 0.3 g/min.

atoms will be formed during oxidation. This may be the reason why the yield of acrolein from 2,2 dimethyl butane was much smaller than that from the other three fuels.

Evolution of Acrolein and CO from Oxygen-Containing Compounds

The yields of acrolein and CO in combustion and/or pyrolysis of n-propyl alcohol, n-amyl alcohol and a 50% sucrose/water solution are shown in Figure 7 through Figure 9, respectively. As Figures 7 (a) and (c) and Figures 8 (a) and (c) show, the maximum yield of acrolein in pyrolysis is much smaller than that in combustion. The maximum

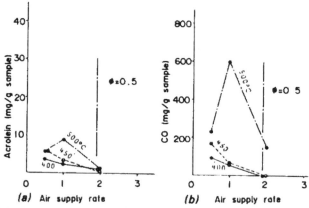

Figure 6. Yields of acrolein and CO from combustion of 2,2 dimethyl butane. Fuel supply rate: 0.3 g/min.

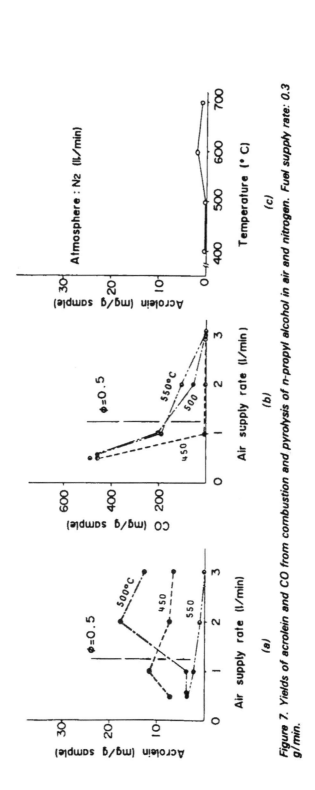

Figure 7. Yields of acrolein and CO from combustion and pyrolysis of n-propyl alcohol in air and nitrogen. Fuel supply rate: 0.3 g/min.

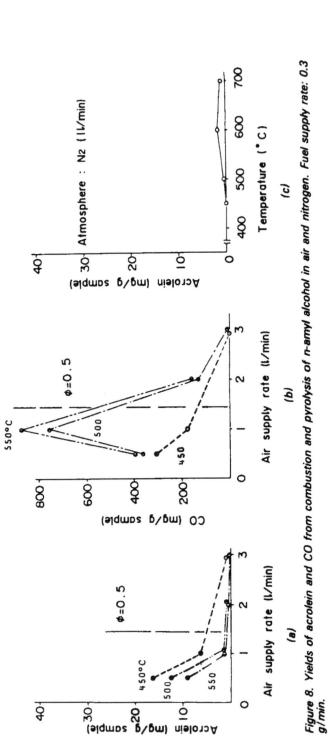

Figure 8. Yields of acrolein and CO from combustion and pyrolysis of n-amyl alcohol in air and nitrogen. Fuel supply rate: 0.3 g/min.

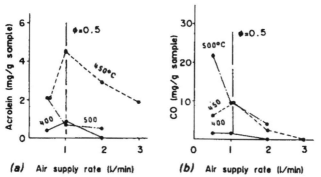

Figure 9. Yields of acrolein and CO from combustion of sucrose. Sucrose: 50% water solution. Fuel supply rate: 0.3 g/min (as sucrose).

yield of CO occurred at an equivalence ratio of fuel to oxygen, ϕ, of approximately 0.25 (see Figures 7(b) and 8(b)), as in the case of the hexanes.

The maximum yields of acrolein and CO from sucrose were far smaller than those of the other two oxygen-containing compounds, as shown in Figures 9 (a) and (b). This may have been due to the formation of a considerable carbonaceous residue in the case of sucrose.

Fuel Mass per Unit Room Space Estimated to be Lethal due to Acrolein or CO under Optimum Burning Condition for Acrolein Evolution

Terrill [2] reported that the lethal concentrations of acrolein and CO for a short time (10 min) exposure are 30 ppm and 5000 ppm, respectively. Using these data, the "lethal fuel mass" per 1 m³ of room space, as a result of acrolein or CO production, was calculated. The burning condition for maximum acrolein production was used for each of the materials plotted in Figure 2 through Figure 6. The results are shown in Figure 10 and Figure 11 as the plots of the minimum burning mass of n-hexane (g/m³) in pure air and nitrogen-diluted air estimated to be lethal due to acrolein against that for CO under the specified burning condition.

The results for the four hexane isomers are shown in Figure 12. It is apparent from these figures that the minimum fuel mass estimated to be lethal due to acrolein evolution is smaller than that for CO under the burning condition for maximum acrolein production. This suggests that in some fires, especially smoldering fires, acrolein may be an important toxicant in addition to CO, depending on the material involved. In Figure 11, the data for the air supply rate of 1 l/min, without nitrogen gas dilution, from Figure 2 are included in the form of

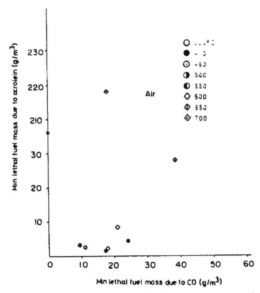

Figure 10. Minimum mass of n-hexane per 1 m³ room space estimated to be lethal due to acrolein or CO under the burning condition for maximum acrolein production. All the data are from Figure 2.

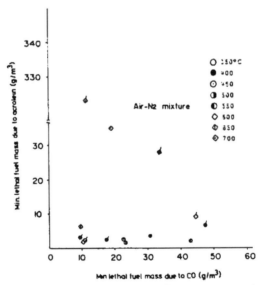

Figure 11. Minimum mass of n-hexane per 1 m³ room space estimated to be lethal due to acrolein or CO under the burning condition in air or nitrogen-diluted air for maximum acrolein condition. All the data are from Figure 3. Symbols with a dash: Values obtained from combustion in pure air.

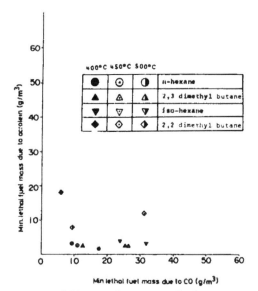

Figure 12. Minimum mass of 4 hexane isomers per 1 m³ room space estimated to be lethal due to acrolein or CO under the burning condition for maximum acrolein production. All the data are from Figures 2, 4, 5 and 6.

the plots with a dash. It can be seen that the minimum burning mass of fuel estimated to be lethal due to acrolein was mostly larger in nitrogen-diluted air than in pure air.

CONCLUSIONS

For fuels used in the present experiments, the results are summarized as follows.

1) Acrolein may have a greater lethal role than CO in the case of combustion at 650 °C or below (generally non-flaming).

2) The suppression of flaming upon dilution of oxygen concentration with nitrogen gas can lead to an increase in the yield of acrolein.

3) The yield of acrolein depends on the molecular structure of the fuels.

4) The maximum yield of acrolein from oxygen-containing materials is much smaller in pyrolysis than in combustion.

REFERENCES

1. Morikawa, T., *Journal of Combustion Toxicology*, Vol. 3, p. 135 (1976).
2. Terrill, J.D., et al, *Science*, Vol. 200, p. 1343 (1978).

Use of a Radiant Furnace Fire Model to Evaluate Acute Toxicity of Smoke

GEORGE V. ALEXEEFF

Toxicologist
Fire Technology Unit
Weyerhaeuser Company
P.O. Box 188
Longivew, Washington 98632

STEVEN C. PACKHAM

Toxicologist and President
BETR Sciences, Inc.
996 So. 1500 East
Salt Lake City, Utah 84105

ABSTRACT

The design and construction of a radiant furnace are described. Acute toxicity experiments were conducted using the radiant furnace as a fire model to produce smoke from Douglas fir (D. fir), southern yellow pine (SYP), and tempered hardboard (THB). The correlation of blood carboxyhemoglobin (COHb) and mortality to irradiation time, smoke concentration, mass loss, carbon monoxide (CO) concentrations, and carbon dioxide (CO_2) concentrations is reported. At 2.5 w/cm², toxicity was characterized by post-exposure lethality with sublethal COHb exposure levels. LC_{50}'s (median lethal concentrations) were 59.5 (D. fir), 66.6 (SYP), and 86.5 mg/L (THB) and the IT_{50}'s (median lethal irradiation times) were 6.8 (D. fir), 6.9 (SYP), and 9.3 minutes (THB). For 5.0 w/cm², most deaths occurred during the 30 minute exposure and COHb levels were in a lethal range. LC_{50}'s at 5.0 w/cm² were 101.6 (D. fir), 100.1 (SYP), and 58.1 mg/L (THB), and the IT_{50}'s were 2.7 (D. fir, SYP) and 2.6 minutes (THB). Data were compared to those produced in other combustion toxicity test methods, and the advantages of the radiant furnace with continuous sample mass-loss monitoring are discussed.

INTRODUCTION

THE TOXICITY OF smoke produced in fires has become a topic of considerable interest to the public, legislative bodies, academia, industry, consumer groups, and fire personnel. In 1982, fires in the United States resulted in 6,020 civilian deaths, 117 fire fighter deaths, 30,525 civilian injuries, and 99,330 fire fighter injuries [1]. Inhalation of smoke is the primary cause of death of fire victims and is a major cause of civilian injury.

Combustion toxicologists typically require the use of at least two models to represent elements of actual fire scenarios, a combustion furnace or fire model in which test materials can be heated and burned and a biological model to determine possible physiological effects of smoke inhalation on humans. This paper deals primarily with the influence of the fire model on the production of toxic gases.

In combustion toxicology, combustion furnaces are used to generate smoke under very controlled conditions. Since the toxicity of smoke from virtually all materials is dependent on thermal and oxidative conditions under which they are degraded, the fire model becomes a significant factor defining both the relevance and the validity of every combustion toxicity test method.

A number of laboratory test methods have been developed in recent years with the intent to obtain data that could be used to evaluate toxic hazard associated with smoke produced from materials in a fire. These methods include combustion devices such as the cup furnace [2-3], tube furnace [4-5], muffle furnace [6] and the radiant furnace. Although most of these test procedures were recently described and summarized [7], there has been little discussion on how the various furnaces affect toxic smoke production, exposure protocols, or toxicologic results. In the opinion of the authors, more research is needed to determine the critical variables essential for both fire and biological models to assure their relevance to one or more fire scenarios. Systematic errors and confounding factors, such as the thermal stability and charring characteristics of the test material, should be minimized or eliminated when possible.

This study considered the production of toxic smoke from three wood products. The results demonstrate how the radiant furnace combined with mass loss measurements can be used to assess a material's thermal characteristics and the effect of those factors on toxicologic observations. We hope that this paper will generate discussion in the field of combustion toxicology and that a closer look will be taken at the various smoke generating systems.

METHODS

Exposure Apparatus

The animal exposure system used in this study consisted of a combustion cell, heat lamps, reflectors, and a 200 L smoke collection exposure chamber as shown in Figure 1. The positioning of the combustion cell with tungsten filament quartz heat lamps (GE QH2M/T3/CL/8T) and reflectors is depicted in Figure 1 and detailed in Figure 2. The lamps had a 254 mm lighted length with a power rating of 2000 w. Incident heat-flux calibrated over a 76 × 177 mm area at a level corresponding to the upper surface of the test sample varied less than 20 percent. Similar calibrations can be conducted on samples up to 38 mm in thickness. The parabolic reflectors were made of stainless steel frames overlaid with a polished aluminum reflective lining. The frame of the combustion cell, shown in Figure 3, was constructed of 23 gauge stainless steel. Quartz glass windows and floor were inserted into the frame and were sealed with heat-resistant gasketing material (1/8" Cerafelt® paper, Johns-Manville). Sample test specimens were held in the combustion cell in a horizontal position on a stainless steel tray which was connected to a load cell for continuous mass monitoring. The combustion cell was positioned under a connecting chimney leading to the constant volume exposure chamber. This design and placement of the combustion chamber was found to be suitable for applying a high heat-flux (up to 7 w/cm²) to a relatively large sample area (up to 76 × 177 mm).

The exposure chamber was constructed of clear polycarbonate with inside dimensons of 1220 × 360 × 460 mm. Inlet ports provided for a head-only inhalation exposure. The animals were restrained in anodized aluminum restrainers as described in Figure 4. The exposure chamber is modeled after the National Bureau of Standards design [3]. For flaming experiments, an electric ignition element was found suit-

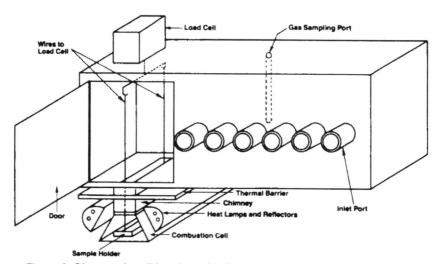

Figure 1. Diagram describing the animal exposure system and radiant furnace.

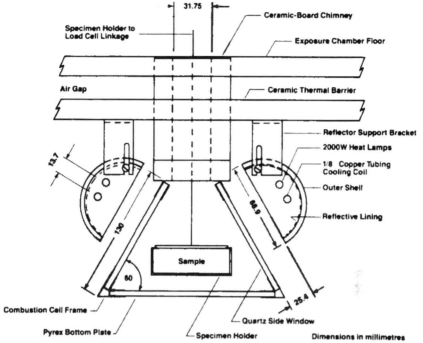

Figure 2. Detailed evaluation showing the radiant furnace assembly.

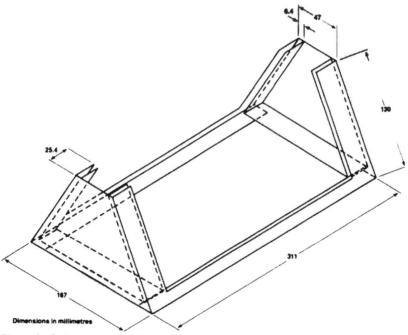

Figure 3. Orthogonal view illustrating the stainless steel combustion cell frame.

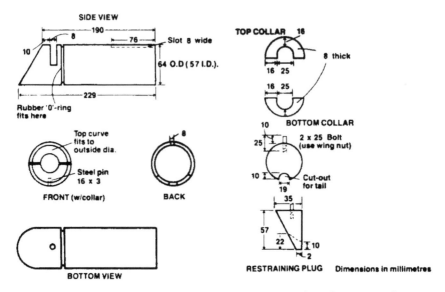

Figure 4. Detailed schematic describing the construction of a rat restrainer.

able to standardize ignition. The element was a coil of 36 gauge nickel chromium wire, 37.5 mm in length, and positioned 10 mm from the edge of the sample. The current through the coil was adjusted by setting a variable transformer to 10%.

Test Specimens

Southern yellow pine (SYP) and tempered hardboard (THB) samples were supplied by the National Forest Products Association and the Douglas fir (D. fir) was purchased from a local commercial supplier. The SYP and D. fir were clear vertical grain studs (38.1 × 88.9 × 2438.4 mm). The THB was a screen-backed sheet (1220 × 2440 × 6 mm). The D. fir and SYP were solid wood products, while THB was constructed of fiberlike wood components with added binding agents and drying oils. It was thought that the additives may influence the toxic smoke properties of the wood product. Test specimens from all three test materials were cut to approximately 76 × 177 × 6 mm and preconditioned at 21°C (70°F) and 50% RH for at least seven days before testing. For mass loss experiments, test specimens were 76 × 76 × 6 mm. Some D. Fir samples, treated with a single coating of intumescent flame retardant (Albi 107A, Albi Manufacturing, East Berlin, Connecticut) were also evaluated for mass loss characteristics.

Chemical Analysis

Concentrations of carbon monoxide (CO), carbon dioxide (CO_2), and oxygen (O_2) in the smoke collection chamber were monitored continuously from sampling ports located in the geometric center of the smoke collection chamber at animal nose level. CO and CO_2 were monitored using an Infrared Industries (Santa Barbara, California) Model #IR-702 dual gas analyzer. Oxygen concentration was monitored with a Sybron/Taylor (England) Servomex oxygen analyzer OA 580. Data was collected using an Apple II Plus micro computer.

Animals

Animals used in the study were F-344 male rats, 275-325 grams, obtained from Simonsen Laboratories (Gilroy, California). The animals were acclimatized on site for at least seven days prior to exposure and were provided food (Ralston Purina's Formula 5001, St. Louis, MO) and water *ad libitum*. One animal in each exposure was used for carboxyhemoglobin analysis. Blood samples were obtained by cardiac puncture using a syringe rinsed with a solution of 100 units sodium heparin/mL saline. Percent COHb was determined using an Instrumentation Laboratory 282-CO-Oximeter (Lexington, MA).

Test Protocol

The wood test sample was cut, weighed, and positioned in the combustion cell. Six animals per test were weighed, restrained, and inserted head-only into the exposure chamber. All instruments were calibrated and zeroed, and the chamber door secured. Each test was begun with the initial delivery of power to the radiant lamps. Test specimens of equal size were exposed to a constant heat-flux of either 2.5 or 5.0 w/cm^2. Smoke concentration was controlled by increasing the irradiation time and thus converting more specimen mass to smoke. When the lamps were turned off, the animals continued to be exposed to smoke for a total exposure time of 30 minutes. At the end of each test, animals were removed from the exposure chamber and the chamber was vented. Blood was taken at this time for carboxyhemoglobin analysis. An animal was prespecified for sacrifice and blood analysis. Such sacrificed animals were not used in the mortality calculations. However, in cases where animals succumbed during the 30-minute exposure, they were used in mortality calculations. The remaining animals were returned to their cages and held for observation for fourteen days. Mortality ratios were used to develop 0 to 100% mortality response curves for each test material. Concentration-response curves with corresponding LC_{50}'s (median lethal concentration) were determined [8,9]. Irradiation time-response curves and cor-

responding IT_{50}, (irradiation times estimated to produce smoke causing 50% mortality) were calculated [8,9]. For mass loss experiments,
samples were irradiated for 30 minutes.

RESULTS

The mean surface areas (cm^2 ± standard deviation) of test specimens
exposed to irradiation were: 134.57 ± 0.15 (D. fir), 134.21 ± 0.87 (SYP),
and 135.06 ± 1.02 (THB). Thus, the area receiving the irradiation was
essentially constant for the three materials. Summaries of toxicologic
and decomposition data for each material as a function of sample irradiation time and maximum exposure concentrations are presented in
Table 1 (D. fir), 2, (SYP), and 3 (THB).

Information on carbon monoxide is presented in each table as the
maximum concentration of CO produced, the average CO concentration per mg/L of total smoke, the CO_2/CO ratio and the percent of carboxyhemoglobin (COHb) determined in whole blood. All parameters
(except CO_2/CO ratio) increased with increasing irradiation time. The
range of irradiation times producing smoke that resulted in mortality
for each individual material was 4.5 minutes at 2.5 w/cm^2 but only 1
minute at 5.0 w/cm^2. In general, carbon monoxide attained potentially
lethal concentrations (4000–5000 ppm for a 30 minute exposure [10]) in
many of the flaming experiments and in a few of the most severe
nonflaming experiments. Maximum CO levels appeared to increase
proportionately with mass loss, however, in all but two experiments,
the total sample mass loss was less than 50%.

The Tables 1–3 contain information on the percent of sample mass

Table 1. Summary of decomposition and toxicologic data for Douglas fir.

Flux (w/cm²)	Maximum Total Smoke (mg/L)	Irradiation Time (minutes)	Maximum CO (ppm × 1000)	Avg. CO (mg/L)	CO₂/CO ratio	Sample Mass Loss (%)	Mortality (%)	COHb (%)
2.5, NF[a]	41.9	6.00	0.8	13.8	6.3	19.5	0	14.6
	47.5	6.42	0.7	13.1	6.2	22.1	17	6.1
	62.9	6.50	2.0	23.8	2.9	31.3	80	35.5
	63.2	7.00	2.8	33.0	2.3	29.3	80	41.3
	77.5	7.50	2.7	25.0	2.5	38.6	60	50.0
5.0, F[b]	64.6	2.00	2.9	37.5	8.9	29.2	0	73.9
	88.3	2.50	3.2	26.9	11.9	41.4	17	58.5
	100.0	2.67	3.6	30.4	11.6	47.0	17	67.0
	106.1	2.75	5.2	32.8	9.0	49.9	83	84.3
	135.2	3.00	7.0	36.0	8.6	62.2	100	73.5

[a]Nonflaming mode.
[b]Flaming mode.

Table 2. Summary of decomposition and toxicologic data for southern yellow pine.

Flux (w/cm²)	Maximum Total Smoke (mg/L)	Irradiation Time (minutes)	Maximum CO (ppm × 1000)	Avg. CO (mg/L)	CO_2/CO ratio	Sample Mass Loss (%)	Mortality (%)	COHb (%)
2.5, NF[a]	42.9	6.42	0.4	6.3	13.3	14.5	0	7.3
	43.2	6.25	0.4	7.0	8.6	14.9	0	NA[c]
	53.4	6.75	0.7	9.5	4.9	18.2	17	0.1
	55.0	6.00	1.2	12.6	4.8	19.1	0	7.4
	65.6	6.50	1.6	18.8	4.1	22.9	60	39.7
	73.7	7.00	2.4	20.9	2.6	27.3	60	46.7
	89.9	7.50	3.2	22.1	3.3	32.7	100	52.0
	107.3	8.50	4.1	25.8	1.6	36.7	100	56.8
5.0, F[b]	71.8	2.33	2.2	23.6	9.6	25.0	0	NA
	78.5	2.25	1.9	14.6	14.4	27.8	0	54.0
	78.6	2.50	2.8	26.6	5.0	28.7	67	NA
	79.3	2.58	2.9	29.7	4.2	28.8	20	57.2
	83.6	2.67	3.8	36.3	5.9	30.9	0	61.7
	97.9	2.83	4.2	32.2	4.7	36.0	60	67.8
	106.9	2.75	5.0	38.7	8.7	39.3	17	NA
	122.3	2.67	6.4	43.8	6.7	42.4	83	87.3
	140.9	3.00	7.2	41.8	6.1	48.9	100	89.5
	159.9	3.25	8.9	44.0	5.2	52.9	100	90.0

[a]Nonflaming mode.
[b]Flaming mode.
[c]Not analyzed.

Table 3. Summary of decomposition and toxicologic data for tempered hardboard.

Flux (w/cm²)	Maximum Total Smoke (mg/L)	Irradiation Time (minutes)	Maximum CO (ppm × 1000)	Avg. CO (mg/L)	CO_2/CO ratio	Sample Mass Loss (%)	Mortality (%)	COHb (%)
2.5, NF[a]	71.4	7.50	1.9	17.8	2.9	16.9	0	31.1
	76.3	8.50	2.2	19.0	2.4	18.0	20	30.9
	90.1	8.75	2.8	18.8	2.3	21.4	60	30.3
	92.3	9.00	3.1	22.2	2.3	22.2	100	51.2
	105.5	10.00	3.7	23.7	2.5	24.8	100	67.0
	114.6	11.00	5.0	26.6	2.0	27.0	100	70.0
	114.8	11.50	4.7	27.4	2.2	27.5	80	70.4
	142.7	12.00	6.3	28.5	2.0	34.7	100	77.6
5.0, NF	49.6	2.50	1.2	17.5	3.2	11.7	40	27.2
	69.7	2.75	2.3	26.9	2.7	16.5	100	49.0
	78.2	3.00	2.6	24.1	2.2	18.5	80	45.4
	81.8	3.25	3.1	26.6	1.7	19.7	80	56.9
	102.7	3.50	4.7	32.2	1.4	24.7	100	53.7

[a]Nonflaming mode.

volatilized during each test and the percent of animal mortality. Concentration-response curves based on these data are plotted for increasing maximum total smoke concentrations for fluxes of 2.5 and 5.0 w/cm² in Figure 5. The calculated LC_{50}, are presented in Table 4. The D. fir and SYP smokes were less toxic (higher exposure concentrations required to produce mortality) at the greater irradiation intensity where flaming occurred. The THB smoke appeared to be more toxic at 5.0 w/cm² even though the sample did not flame.

In Figure 6, the irradiation time-response curves for each material are plotted for increasing irradiation times at 2.5 and 5.0 w/cm². The calculated IT_{50}, are presented in Table 5. A shorter irradiation time was required to produce smoke causing mortality (toxic smoke) when the higher heat flux was used. THB required a longer irradiation time to produce toxic smoke at 2.5 w/cm², but all three materials required the same length of time to produce toxic smoke at 5.0 w/cm².

Table 6 compares CO and CO_2 generation in three smoke toxicity

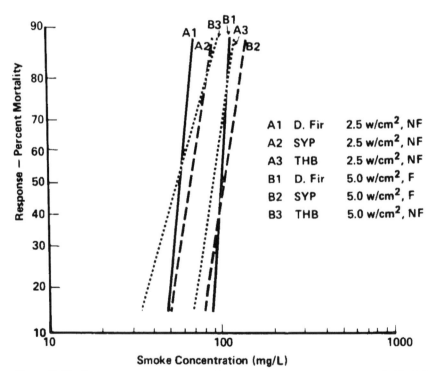

Figure 5. *Maximum exposure concentration-response curves for Douglas fir (D. fir), southern yellow pine (SYP), and tempered hardboard (THB) using the radiant furnace. The materials were tested at two heat fluxes: 2.5 w/cm² and 5.0 w/cm². The samples were either nonflaming (NF) or flaming (F) during the test.*

Table 4. LC_{50} values determined.

Material	Flux (w/cm²)	Mode	LC_{50} (mg/L)	95% CI
Douglas fir	2.5	NF[a]	59.5	50.9- 75.5
	5.0	F[b]	101.6	92.0-119.2
Southern yellow pine	2.5	NF	66.6	60.5- 75.5
	5.0	F	100.1	90.5-112.9
Tempered Hardboard	2.5	NF	86.5	79.4- 93.8
	5.0	NF	58.1	40.8- 67.8

[a]NF refers to nonflaming conditions.
[b]F refers to flaming conditions.

Table 5. IT_{50} values determined.

Material	Flux (w/cm²)	Mode	IT_{50} (Minutes)	95% CI
Douglas fir	2.5	NF[a]	6.8	4.8-11.5
	5.0	F[b]	2.7	2.6- 2.8
Southern yellow pine	2.5	NF	6.9	6.7- 7.2
	5.0	F	2.7	2.6- 2.8
Tempered Hardboard	2.5	NF	9.3	8.7- 9.9
	5.0	NF	2.6	2.0- 2.8

[a]NF refers to nonflaming conditions.
[b]F refers to flaming conditions.

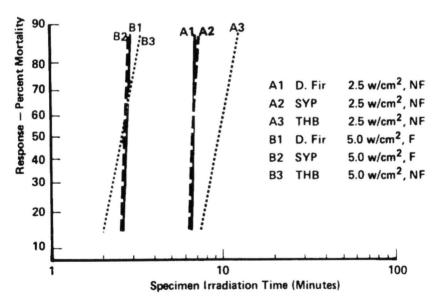

Figure 6. Specimen irradiation time-response curves for Douglas fir (D. fir), southern yellow pine (SYP), and tempered hardboard (THB) using the radiant furnace. The materials were tested at two heat fluxes: 2.5 w/cm² and 5.0 w/cm². The samples were either nonflaming (NF) or flaming (F) during the test.

Table 6. Comparison of CO and CO_2 generation in the cup, radiant, and Pittsburg furnaces using Douglas fir.

Furnace	Mean of: Avg. ppm CO Produced/ (mg/L) of Total Smoke	Mean of: Avg. ppm CO_2 Produced/ (mg/L) of Total Smoke	CO_2/CO Ratio
Cup[a], NF	110 ± 19	260 ± 88	2.4
	76 ± 6.7	690 ± 83	9.1
Radiant			
2.5 w/cm², NF	21 ± 9.4	72 ± 19.5	3.4
5 w/cm³, F	33 ± 4.3	344 ± 33.8	10.4
Pittsburg[b], F	25	114[c]	4.6

[a]Levin et al. 1983.
[b]Alarie and Anderson, 1979.
[c]Based on maximum reported CO_2 in Alarie and Anderson, 1979.
[b,c]Data converted to mg/L with the assumption that 63.5g were mixed with a 20L air flow for 30 minutes.

test methods. Data were converted to average ppm CO produced per mg/L of total smoke to allow comparison with NBS data published on the cup furnace. The data for the Pittsburg-muffle furnace is an estimate from Alarie and Anderson [11]. The radiant and Pittsburg-muffle furnaces appear to be in close agreement while the cup furnace produced substantially higher levels of CO and CO_2.

Percent mass loss over time of 76 × 76 mm samples of D. fir, SYP, and THB are presented in Figure 7. Data from a flame-retardant treated sample of D. fir showed a substantial decrease in rate of mass loss compared to the other wood samples. Total mass lost from the FR treated D. fir was also reduced in comparison to untreated test specimens. The average time-weighted CO concentrations produced in these experiments were 5155 (untreated) and 795 (treated) ppm. The mean ppm CO/(mg/L) ratios were 69.0 (untreated) and 18.3 (treated).

SUMMARY AND DISCUSSION

Acute toxicity experiments were conducted using a radiant heat source as a fire model to produce smoke from three wood products, D. fir, SYP, and THB. Smoke concentration was controlled by exposing sample specimens of equal size to a constant heat flux for varying periods of time. As shown in Figure 7, the specimen mass loss was converted to smoke as the specimen irradiation time was increased. In practice, radiant energy was applied for a range of time intervals until a 0 to 100% mortality response curve could be developed for each test material. Results were expressed in terms of both the calculated specimen irradiation time required to produce smoke estimated to cause 50% mortality (IT_{50}) and the calculated mass lost per volume of exposure chamber volume estimated to cause 50% mortality (LC_{50}). The mass of the test specimen was continuously monitored to provide information on the rate of smoke production.

The toxicity of D. fir and SYP smoke was less under flaming than under nonflaming conditions (flaming LC_{50}'s were 101.6 and 100.1; nonflaming LC_{50}'s were 59.5 and 66.6 for D. fir and SYP respectively). The qualitative nature of the toxic effect was also different under the two combustion conditions. Post-exposure respiratory symptoms leading to death were common in subjects exposed to nonflaming wood smoke. Carboxyhemoglobin levels were below 55% in these subjects at the termination of exposure and were insufficient to cause death. In contrast, animals exposed to flaming combustion products consistantly exhibited COHb levels above 50% and some as high as 80% suggesting a strong CO intoxication factor.

The toxicity of THB combustion products was distinct from that of D. fir and SYP. THB smoke toxicity increased with increasing heat flux from 2.5 to 5.0 w/cm². Since flaming did not occur at 5.0 w/cm², the shift in LC_{50} cannot be attributed to differences in the mode of combustion. The apparent increase in smoke toxicity may have resulted from a longer exposure time to the combustion products due to a more rapid

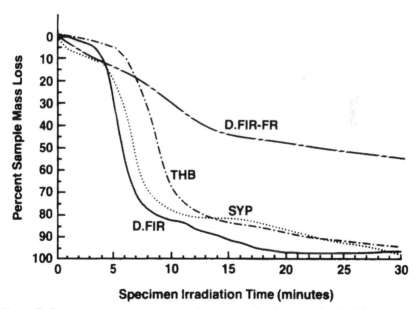

Figure 7. Percent mass loss-concentration curves for Douglas fir (D. fir), southern yellow pine (SYP), and tempered hardboard (THB) and a flame retardant sample of Douglas fir (D. fir-FR) using the radiant furnace. The materials were tested at 2.5 w/cm² under nonflaming conditions.

rate of mass loss. The CO intoxication appeared to be a greater factor at 2.5 w/cm^2 than at 5.0 w/cm^2 opposite to the trend exhibited by animals exposed to D. fir and SYP smoke, thus new combustion products may also have been formed.

Comparisons can be made between the current data for D. fir and those reported for the NBS cup [3] and the Pittsburg methods [6]. The LC$_{50}$'s under nonflaming conditions for the NBS and radiant methods differ by a factor of 2.6 (radiant = 59.5 mg/L; NBS = 22.8 mg/L). Under flaming conditions, they differ by a factor of 2.8 (radiant = 101.1 mg/L; NBS = 36.0 mg/L). One explanation for this difference may lie in the fact that in the radiant tests less than 50% of the original test specimen mass was volatilized, while in the cup test [2] over 95% mass loss occurred for D. fir. In the authors' opinion, this implies that the toxic potency of smoke from the first 50% of mass lost from a specimen of wood may be less than that produced from combustion of the second 50%. More detailed experiments would be required to adequately test this hypothesis, but this study raises an interesting issue in that materials may not release toxicants of equal potency through progressive stages of thermal degradation. If this trend is verified, toxic hazard evaluation and modeling strategies may have to be altered.

Alarie and Anderson [6,11] reported an LC$_{50}$ of 63.8 grams for D. fir under conditions of ramped temperature profiles unique to the Pittsburg method. On the basis of a 20 L/min flow rate used in their flow through system and the time period over which sample mass loss occurred an average exposure concentration of 106 mg/L can be calculated assuming a 63.8 g sample. This concentration is almost identical to the LC$_{50}$ for D. fir under flaming conditions in the radiant furnace as reported here. Calculated estimates based on these same published data [6,11] further indicate that the amount of CO and CO$_2$ produced per mg/L of total smoke is also similar for the Pittsburg and radiant fire models (Table 6). The cup furnace, for which LC$_{50}$ values were reported to be one-third to one-fourth as high as those of the Pittsburg and radiant methods, also produced greater quantities of CO and CO$_2$ per gram of D. fir. CO yields may be one of the reasons for the observed differences between toxicologic results for wood in these respective methods.

The tests conducted at a fixed heat flux on all three wood products plus flame retardant treated D. fir indicated that the rate of mass loss was material dependent (Figure 7). Since dosage is proportionate to the product of smoke concentration and exposure duration, it is readily apparent that different mass loss rates will result in different dose levels even when the exposure time and specimen sizes are equal. This inherent tendency for a material's thermal characteristics to influence actual dosage is frequently overlooked in combustion toxicity testing. The possible magnitude of such an oversight is clearly demonstrated here. Forty-nine percent of the treated D. fir sample specimen remained after 30 minutes while less than 5% of the untreated sample remained when both specimens were exposed to 2.5 w/cm^2. This would represent an error of 94% in this particular case if exposure concentration calculations did not take into account actual mass loss, but had relied on mass loaded into the furnace (as was the case in the NBS study).

Sample mass loss over time causes exposure concentrations to change throughout smoke toxicity experiments when constant volume chambers are used. In most cases, however, concentration values which are reported reflect only maximum exposure concentrations which may not have been obtained until near the end of the exposure period. It would seem advisable to have a standard expression of dosage based on actual smoke concentration (mg/L) and exposure time (minutes). If this were done, then dosage expressed as the product of (mg/L) and minutes could be determined for both constant volume and flow through systems. Results from different test procedures could be normalized and meaningful comparisons made. Measuring test specimen mass throughout the experiment would be essential if a normalization of exposure dosage was desired. Clearly, the ability to calculate more precise dosage values (such as mg/L times minutes) and the analysis of toxicant production as a function of mass loss are important for precise assessment of relative toxicity of smokes.

In addition to LC_{50}'s based on actual mass lost from the test specimen, the radiant furnace procedure provided information on when toxic smoke was produced. Since the median lethal irradiation time (IT_{50}) expresses the time required under specific heat-flux conditions for toxic smoke to be released following initial application of radiant energy, it may be useful in separating thermal properties of materials from properties of toxic potency.

The radiant system introduced in these studies exhibits a functional compatibility with the NBS 200 L chamber and a flexibility in operation which in the authors' opinion can greatly advance combustion toxicity research and testing. The recording of mass loss over time can provide for a more meaningful expression of dosage in mg/L times minutes. Variation in irradiation duration and intensity can be used to help prevent properties of thermal stability from confounding toxic potency evaluations.

This study indicates the greater complexity of variables involved in combustion toxicity testing in comparison to single gas inhalation exposures. The radiant furnace method's versatility makes it an ideal candidate for further research in smoke toxicity. However, due to the lack of understanding of the importance and influence of a number of testing procedure variables, a single furnace method cannot be supported for the regulation of products on the basis of smoke toxicity at this time.

ACKNOWLEDGEMENTS

The authors acknowledge the technical expertise of Howard Stacy, presently at Southwest Research Institute, in the development of the radiant furnace. We wish to thank our secretary Penny Petta for her assistance in the preparation of this paper. We also acknowledge the National Forest Products Association for their support of this project.

REFERENCES

1. Karter, M. J., Jr., "Fire Loss in the United States During 1982," *Fire J.* 77: 44 (1983).
2. Potts, W. J., and Lederer, T. S., "A Method for Comparative Testing of Smoke Toxicity," *J. Comb. Toxicol.* 4: 114-162 (1977).
3. Levin, B. C., Fowell, A. J., Birky, M. M., Paabo, M., Stale, A., and Malek, D., "Further Development of a Test Method for the Assessment of the Acute Inhalation Toxicity of Combustion Products," National Bureau of Standards (U.S.), NBSIR82-2532 (1982).
4. Klimisch, H. J., Hollander, H. W., and Thyssen, J., "Generation of Constant Concentrations of Thermal Decomposition Products in Inhalation Chambers. A Comparative Study with a Method According to DIN 53-436. 1. Measurement of Carbon Monoxide and Carbon Dioxide in Inhalation Chambers," *J. Comb. Toxicol.* 7: 243-256 (1980).
5. Hilado, C. J., "Evaluation of the NASA Animal Exposure Chamber as a Potential Chamber for Fire Toxicity Screening Test," *J. Comb. Toxicol.* 2: 298-314 (1975).
6. Alarie, Y. C., and Anderson, R. C. "Toxicologic Classification of Thermal Decomposition Products of Synthetic and Natural Polymers," *Toxicol. Appl. Pharmacol.* 57: 181-183 (1981).
7. Kaplan, H. L., Grand, G. F., and Hartzell, G. E., *Combustion Toxicology: Principles and Test Methods*, Technomic Pub. Co., Inc., Lancaster, PA (1983).
8. Finney, D. J., *Probit Analysis*, Third Edition, Cambridge University Press, Cambridge, England (1971).
9. Lieberman, H. R., "Estimating LC_{50} using the Probit Technique: A Computer Program," *Drug Chem. Toxicol.* 4: 114-162 (1983).
10. Kimerle, G., "Aspects and Methodology for the Evaluation of Toxicological Parameters During Fire Exposure," *JFF/Comb. Toxicol.* 1: 4-50 (1974).
11. Alarie, Y. C., and Anderson, R. C., "Toxicologic and Acute Lethal Hazard Evaluation of Thermal Decomposition Products of Synthetic and Natural Polymers," *Toxicol. Appl. Pharmacol.* 51: 341-362 (1979).

Modeling of Toxicological Effects of Fire Gases: I. Incapacitating Effects of Narcotic Fire Gases

HAROLD L. KAPLAN and GORDON E. HARTZELL

Department of Fire Technology
Southwest Research Institute
Postal Drawer 28510
San Antonio, Texas 78284

ABSTRACT

Common fire gas toxicants fall into two major classes, narcosis-producing agents and irritants. It is desirable to model, mathematically, the effects of these common toxicants on humans exposed in a fire, and therefore, obviate the use of large numbers of laboratory animals in smoke toxicity testing. From a review of methodologies for assessment of the incapacitating effects of the narcotic fire gases, it appears that rats are sensitive to approximately the same range of accumulated doses as may be deemed potentially hazardous to human subjects. This paper introduces an approach as a first approximation to the modeling of the incapacitating effects of the narcotic toxicants based on correlation of observed effects with accumulated doses *to which subjects are exposed.*

INTRODUCTION

IN ALL FIRES, smoke is produced, which consists of a complex mixture of solid particulates, liquid aerosols and gaseous products. Over 100 chemical species have been identified from the burning of some materials [1]. Although combustion may generate a wide variety of products, the common toxicant gases may be separated into two major classes, the narcotic agents and the irritants. In pharmacologic terminology, "narcotic" is defined as a drug which induces unconsciousness and relieves pain by its direct action on the central nervous system [2]. In combustion toxicology, the term is used to refer to those toxicants which are capable of producing a loss of consciousness by either direct or indirect effects on the central nervous system. In the case of some of these toxicants, the indirect effects result from their asphyxiant action. Many narcotic agents may be produced by the combustion of materials, including carbon monoxide, carbon dioxide, hydrogen cyanide, benzene, acetone, toluene and various alcohols and aliphatic hydrocarbons. However, only carbon monoxide, carbon dioxide and hydrogen cyanide have been measured in smoke in sufficient concentrations to cause significant acute toxic effects. In addition, the condition of oxygen vitiation may also cause unconsciousness and some consider oxygen depletion as a narcotic.

Although, in sufficient concentration, a narcotic chemical will produce a loss of consciousness and even death, the effects of these agents at much lower concentrations may be particularly important in a fire environment. At low concentrations, these substances may act on the central and peripheral nervous systems, resulting in impairment of sensory, mental and/or motor functions [3]. Such impairment, depending on its extent, may cause delay or prevention of escape from the fire environment, with resultant death due to continued inhalation of toxic gases, thermal burns and other factors.

In the development of laboratory combustion toxicity tests, the potential importance of the sublethal effects of narcotic combustion products has long been recognized [4]. However, in recent years, the primary emphasis of most laboratory test methods has still been largely on lethal effects. This is not surprising in view of the fact that the end point for assessing the relative toxicity of chemicals has traditionally been lethality. Additionally, serious concerns have been raised regarding the relevance of sublethal rodent effects to human subjects. In spite of these concerns, continued studies of sublethal effects of smoke have led to the development and application of a variety of "behavioral" methodologies to assess the escape-impairing effects of combustion atmospheres. These effects have been termed "incapacitation," "impairment," "intoxication" and "sublethal effects" by different investigators. Although the methods differ in the endpoints selected, the objective of each is the assessment of the potential of smoke to impair escape. In addition to differences in endpoints, the methods also differ in equipment/instrumentation requirements and in the extent of training required for test animals.

The objective here is to review the major methodologies which have been used to assess the sublethal escape-impairing effects of two primary narcotic combustion gases, carbon monoxide (CO) and hydrogen cyanide (HCN), in rodents. This review includes both a survey of the methods used in combustion toxicity tests and an evaluation of the relevance of each method as a model for predicting effects of CO and HCN-containing combustion atmospheres on humans.

In order to evaluate the predictive capability of each test method for human exposures, it is necessary to relate rodent incapacitation data with existing human and nonhuman primate toxicity data. A number

of difficulties are encountered in making these comparisons. One difficulty results from the use of different experimental protocols by
various investigators and from differences in the forms of data
reported. Few investigators have determined incapacitation times at
several concentrations of narcotic gases, thereby preventing construction of useful concentration-time (Ct) curves [5]. Moreover, most have
reported EC_{50} values for incapacitation corresponding to only one exposure time, e.g., 30 minutes. Therefore, data have been converted here
to Ct products whenever possible in order to obtain a common
denominator to allow comparisons. This treatment of data is based on
Haber's rule, the principle of which is that, within certain time and
concentration ranges, the product of the concentration and the exposure time is a constant, characteristic of the toxicant [6]. The
validity of Haber's rule in the curved portion of the hyperbolic concentration-time curve has been well demonstrated for CO. Perhaps for-
tuitously, this range of the Ct curve corresponds to the range of CO
concentrations and times of particular concern in fires. Haber's rule
also appears to be applicable to incapacitation by HCN, although
limited experimental data indicate that the Ct product may be constant over a much narrower range of the concentration-time curve than
with CO. It should be noted that Haber's rule loses validity as the
asymptotes of the Ct curve is approached. Unfortunately, many investigators have reported data, such as EC_{50} values for 30 minutes,
which lie near the asymptotes of the Ct curves. The Ct products obtained from these data are unusually high and must be interpreted
with caution.

Another problem in evaluating the validity of test methods in
predicting the escape-impairing potential of narcotic fire gases in
humans is the limited availability of relevant human data, particularly
in the case of HCN. Consequently, it has been necessary to use some
subjective judgement in estimating Ct values for escape impairment of
humans. Rationale for these estimates will be discussed further.

INCAPACITATION OF HUMANS

It is well established that the primary toxicant involved in most fire
fatalities due to smoke inhalation is CO [7,8]. The toxicity of CO is
primarily due to its affinity for the hemoglobin in blood. Even partial
conversion of hemoglobin to carboxyhemoglobin (COHb) reduces the
oxygen-transport capability of the blood, thereby resulting in a
decreased supply of oxygen to critical body organs. In addition, CO
causes hemoglobin to cling more tightly to its oxygen content than
would otherwise occur, further decreasing the availability of oxygen to
body tissues. Symptoms observed in humans at various COHb blood
saturation percentages are shown in Table 1. These data indicate that,
at COHb levels below 30%, effects are normally not sufficiently severe

Table 1. Human responses to various concentrations of carboxyhemoglobin [9].

COHb Concentration	Symptoms in Humans
0–10	None
10–20	Tension in forehead, dilation of skin vessels
20–30	Headache, pulsation in sides of head
30–40	Severe headache, ennui, dizziness, weakening of eyesight, nausea, vomiting, prostration
40–50	Same as above, increase in breathing rate and pulse, asphyxiation and prostration
50–60	Same as above, coma, convulsions, Cheyne-Stokes respiration
60–70	Coma, convulsions, weak respiration and pulse, death possible
70–80	Slowing and stopping of respiration, death within hours
80–90	Death in less than an hour
90–100	Death within a few minutes

as to impair human escape capability and, at levels above 40%, escape may be severely impaired or even not possible. Thus, the threshold COHb blood level for escape impairment of human subjects appears to be in the range of 30 to 40%. Concentration-time curves for COHb loading by man with a respiratory minute volume (RMV) of 20 L/min, derived from the Stewart-Peterson equation [10], are shown in Figure 1. From the curves corresponding to 30 to 40 percent COHb, escape impairment of humans engaged in light physical activity would be anticipated after inhalation of an accumulated Ct dose of CO in the range of approximately 35,000 to 45,000 ppm-minutes. Under conditions of strenuous physical activity, the RMV could increase twofold or more, resulting in a more rapid loading of COHb and in incapacitation at a much smaller accumulated Ct dose.

Results of nonhuman primate studies indicate that incapacitating doses of CO in these animals are reasonably comparable to the estimated doses for incapacitation of humans. Studies of the effects of CO on escape capability of the juvenile baboon have been conducted at Southwest Research Institute [11]. The EC_{50} concentration for escape-impairment (shuttlebox paradigm) after a five-minute exposure to CO was determined for the juvenile baboon to be 6850 ppm (Figure 2). This equates to a Ct product of 34,250 ppm-minutes, which is slightly lower than that predicted for human subjects, reflecting differences in activity, RMV and body weight between the two species. Other studies with cynomolgus monkeys conducted at The Huntingdon

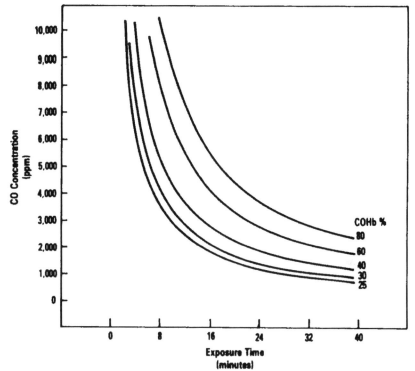

Figure 1. Percent COHb as a function of ambient CO concentration and exposure time for humans with a RMV of 20 L/min. Based on Stewart-Peterson Equation: Log (Δ% COHb/L) = 1.036 Log (CO Concentration) − 4.4793 (Reference 10).

Research Centre report incapacitating CO doses of from 21,000 to 30,000 ppm-min for active animals [12] and of from 45,000 to 55,000 ppm-min for restrained animals [13]. The Ct products for active animals are somewhat lower than the range predicted for human escape impairment, probably due to differences between the two species in activity, RMV and body weight.

Hydrogen cyanide is a very rapidly acting toxicant. Although cyanide ions do not combine appreciably with hemoglobin, they bind with the trivalent ion of cytochrome oxidase in cellular mitochondria, thereby inhibiting the utilization of molecular oxygen by cells. Data relating symptoms in humans to various concentrations of HCN are very limited. The most widely used source of descriptive accounts of human cyanide intoxication is from Kimmerle [9]. Based on these data shown in Table 2, 50 ppm may be tolerated by man for 30 to 60 minutes without difficulty but 100 ppm for that same period is likely to be fatal, 135 ppm may be fatal after 30 minutes and 181 ppm may be fatal

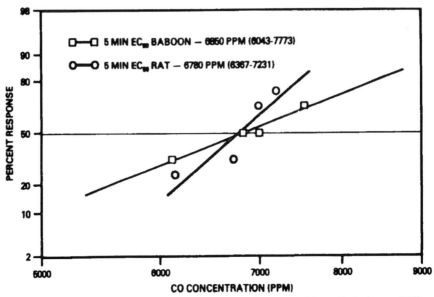

Figure 2. Effects of CO on escape impairment in rats and juvenile baboons [11].

after 10 minutes. If the assumption is made that incapacitation occurs at one-third to one-half the lethal accumulated dose (which appears to be a valid relationship for CO in man and in rats and for HCN in rats), the data suggest that accumulated doses for incapacitation by HCN range from approximately 2,500 ppm-min (100 ppm) to 750 ppm-min (200 ppm). It also appears that concentrations greater than approx-

Table 2. Relation of hydrogen cyanide concentration in air and symptoms of humans [9].

HCN Concentration ppm	Symptoms
0.2-5.1	Threshold of odor.
10	ACGIH TLV-C value
18-36	Slight symptoms (headache) after several hours.
45-54	Tolerated for 1/2 to 1 hour without difficulty
100	Death after 1 hour.
110-135	Fatal after 1/2 to 1 hour, or dangerous to life.
135	Fatal after 30 minutes.
181	Fatal after 10 minutes.
280	Immediately fatal.

imately 200 ppm lie in the asymptotic portion of the Ct curve and accumulated doses necessary to produce incapacitation are reduced considerably.

Nonhuman primate studies of the incapacitating effects of HCN are very limited, the sole source of data being The Huntingdon Research Centre. In restrained cynomolgus monkeys exposed to from 100 to 156 ppm, Ct products at incapacitation decreased with increasing concentrations and ranged from 1900 to 1176 ppm-min [14]. These values appear reasonably comparable to those predicted from Kimmerle's data for human escape impairment.

METHODS FOR MEASUREMENT OF SUBLETHAL EFFECTS ON RODENTS

Leg-Flexion Shock Avoidance

One of the most widely used behavioral models in the laboratory is the leg-flexion shock avoidance response. This method, using rats, was originally applied to smoke toxicity research at the University of Utah [15,16]. Although eventually dropped from the most recent version of the U.S. National Bureau of Standards smoke toxicity test method, the leg-flexion response was used extensively during the development of that test and considerable data were generated [17].

In the leg-flexion shock avoidance response method, the rat is positioned in a tubular restrainer such that an electric shock (ca. 4.5 ma) is delivered to one hind leg upon contact with a suspended metal platform. Approximately 10–15 minutes of training are required to train the rat to avoid the shock by raising the leg above the metal plate. Upon exposure to a toxicant atmosphere, incapacitation is said to occur when the rat is unable to avoid the shock. In most smoke toxicity testing, up to six rats are instrumented and exposed simultaneously. Responses are monitored both visually and also using a recording device. Although the avoidance loss endpoint is somewhat subjective, experience and skill on the part of the observer result in quite reproducible data.

Much of the data obtained from the leg-flexion shock avoidance technique has been expressed as plots of time-to-incapacitation against concentration of toxicant atmosphere, i.e., Ct plots [6]. From these plots, quantification of incapacitating doses for both of the narcosis-producing toxicants, carbon monoxide and hydrogen cyanide, are possible.

Carbon Monoxide

A composite Ct plot for carbon monoxide using the leg-flexion shock avoidance response is shown in Figure 3. The plot, drawn both from

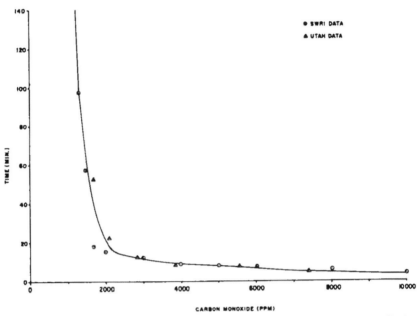

Figure 3. Concentration-time (Ct) curve for incapacitation measured by leg-flexion shock avoidance method in the rat.

early work at the University of Utah (1976) and later studies at Southwest Research Institute (1982), shows the data to be quite comparable [5,18]. From the Ct plot, it can be seen that incapacitation of rats occurs with accumulated CO doses of about 30,000 to 40,000 ppm-min in the curved portion of the Ct curve. These doses result in blood carboxyhemoglobin saturations of about 60 percent for rats at incapacitation.

The leg-flexion response has been criticized as being a reflex action which is too close to death to be useful [17]. However, non-human primate studies at SwRI have shown that the accumulated dose of CO (approximately 34,000 ppm-min) which prevents escape of baboons in a shuttlebox paradigm after a five-minute exposure does indeed relate quite well to the accumulated dose causing loss of the leg-flexion response in rats [12]. Other studies with cynomolgus monkeys report incapacitating CO doses of approximately 25,000 and 50,000 ppm-min, depending upon whether the monkey is active or restrained, respectively [12,13]. From these primate studies, along with data on human experience, it would appear that the rat is a reasonably good model for incapacitation of humans by CO in terms of time-concentration parameters.

Hydrogen Cyanide

The leg-flexion shock avoidance model yields relatively sharp end-points when used for exposure of rats to HCN. An experimental Ct curve for HCN exposures conducted at SwRI is shown in Figure 4 [12]. In the curved portion of the Ct curve (from approximately 130 to 250 ppm), incapacitation of rats occurs with accumulated doses in the range of from about 2700 to 1200 ppm-min, with the Ct products decreasing with increasing HCN concentrations. With estimates of incapacitating doses of HCN for humans probably being in the range of from 2500 to 750 ppm-min, it would appear that the rat is a reasonable model for predicting escape impairment of human subjects exposed to HCN as well as to CO.

Motor-Driven Activity Wheel

The motor-driven activity wheel (tumble cage) has been used in

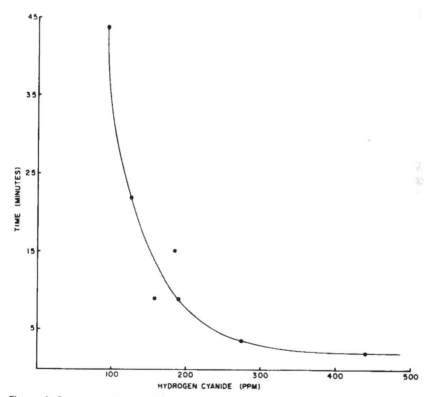

Figure 4. Concentration-time (Ct) curve for effect of HCN on incapacitation by leg-flexion shock avoidance method in the rat [12].

several combustion toxicity test methods to assess the incapacitating effects of combustion atmospheres. Studies include those at the U.S. Federal Aviation Administration, Harvard Medical School, McDonnel Douglas Corporation, U.S. Testing Company and also a reported Japanese test method [19]. Although the size and configuration of the equipment used in these methods vary, the activity wheel method basically consists of a wire-mesh cage in the form of a wheel, which is rotated at a slow constant speed by an electric motor. The animal instinctively moves in a direction counter to that of wheel rotation in an attempt to remain upright in the cage. The animal is considered incapacitated when it is no longer capable of maintaining its movement and begins to slide or tumble as the cage is rotated.

Carbon Monoxide

Investigators at the U.S. Federal Aviation Administration (FAA) used a motor-driven three-compartmented wheel to establish dose-response relationships for incapacitation of rats exposed to CO [20]. The incapacitating dose determined for this gas, expressed as the weight of gas that must be inhaled per kilogram of body weight to produce the effect, was 15.35 mg/kg. From the data, the investigators derived a formula to predict the time required for a concentration of CO to incapacitate a rat of a given weight in their test system. The predicted Ct value for incapacitation of a 300-g rat after a 5 to 30-minute exposure to CO is approximately 37,000 ppm-min, which is within the range of Ct products obtained by the leg-flexion shock avoidance method and the range predicted for human escape impairment. The value is also similar to the Ct product for incapacitation of active baboons, but is somewhat higher than the range of values reported for active monkeys.

The activity/tumble cage was also used to measure the incapacitating effects of CO in rats at the University of Michigan [21]. The tumble cage consisted of four compartments mounted on a shaft coupled externally to a motor which rotated the cage at 6 ± 0.3 rpm. Animals were trained for several sessions prior to testing in order to ensure uniform walking behavior. Four parameters (lag, turn, slide and tumble) were scored during each 1-minute interval during the 30-minute exposure; however, only sliding and tumbling of animals were used as indicators of incapacitation. The onset of repeated or continuous sliding and/or tumbling was recorded as incapacitation. A concentration-response curve and a 30-minute EC_{50} value of approximately 1025 ppm CO for incapacitation were obtained. The Ct product at the EC_{50} value for incapacitation in this system was 30,750 ppm-minutes. This value relates reasonably well to those values reported for the tumble cage by FAA and for the leg-flexion shock avoidance method and to human data.

The activity wheel was used at the McDonnel Douglas Corporation to measure the time-to-incapacitation in rats exposed to CO [22]. The animals were exposed individually in a cage rotated at 4 rpm to CO at average concentrations of from 11,250 to 14,600 ppm. Times-to-incapacitation in the range from 112 to 156 seconds were reported. The Ct products from individual animal data ranged from 22,000 to 36,000 ppm-minutes. Although most values exceeded 30,000 ppm-minutes and were reasonably close to the estimated human escape impairment range, some were considerably lower than this range. However, it should be noted that the concentrations and times for these exposures are not in the curved part of the concentration-time curve.

Hydrogen Cyanide

The motor-driven activity wheel was also used at the U.S. Federal Aviation Administration to investigate dose-response relationships for incapacitation of rats exposed to HCN [20]. The incapacitating dose determined for this gas, expressed as the weight of gas that must be inhaled per kilogram of body weight to produce the effect, was 0.32 mg/kg. Using the data, a formula was derived to predict the time required for a concentration of HCN to incapacitate a rat of a given weight in their test system. The predicted Ct value for incapacitation of a 300-g rat after a 5- to 30-minute exposure to HCN is approximately 800 ppm-minutes, which is reasonably close to the Ct products obtained at the higher HCN concentrations in limited nonhuman primate studies. This value is also close to the Ct product estimated for escape impairment of human subjects exposed to concentrations near 200 ppm.

Pole-Climb Conditioned Avoidance/Escape Response Test

The pole-climb conditioned avoidance/escape response test was developed at SRI International [23]. The test utilized a multisensory conditioned avoidance response (CAR) to any of three sensory stimuli (visual, auditory and tactile) in order to evaluate sensory, perceptual and/or motor deficits during exposure to combustion atmospheres or gases. The apparatus consists of a stainless steel chamber, with an electrified metal floor grid to provide a shock and an aluminum pole suspended from the center of the chamber ceiling. Visual or auditory stimuli provided by a light or loudspeaker mounted on the ceiling are given 10 seconds prior to initiation of a 30-second shock. The rat learns to avoid or escape the shock by climbing onto the pole which is lubricated to discourage the animal from remaining on it. The test system is computer automated and downward displacement of the pole by the rat closes a microswitch, recording the time of the animal's response.

This test system allows examination of two principal parameters. These are the percentage of trials during which each rat avoided shock by responding within the 10-second stimulus interval that preceded shock and the percentage of trials on which each rat failed to respond throughout the trial, including the 20-second shock interval. According to the investigators, avoidance failures are thought to reflect disorientation, sensory deficits and/or slight motor impairment. Escape failures reflect incapacitation under the conditions of the experiments. From the cumulative data obtained during 30-minute exposures, two endpoints are calculated: (1) the concentration of combusted material or gas that caused a 50% reduction in the CAR performance compared with controls (CC_{50}) and (2) the concentration of combusted material or gas that caused a 50% failure to respond to the shock (IC_{50}). In addition, the time to reach the first end point (a 50% CAR decrement) is determined and is called the Time to Avoidance Deficit (TAD).

Carbon Monoxide

In experiments conducted with pure CO, rats were exposed for 30 minutes to the gas at concentrations ranging from 1000 to 3240 ppm. At the highest concentration tested, 3240 ppm, a TAD of 9.6 minutes was reported. The Ct value for this exposure is 31,104 ppm-min, which is reasonably close to the estimated 35,000 to 45,000 ppm-min for escape impairment in humans and to Ct products obtained with other rodent methods. In another publication by this laboratory, the EC_{50} values for avoidance and escape deficits after a 30-minute exposure to CO were reported as 1600 and 3125 ppm, respectively [24]. The Ct value of 48,000 ppm-min for avoidance deficit is close to the range estimated for escape impairment of humans but the Ct value of 93,750 ppm-min for escape deficit is considerably higher than this range and the Ct products obtained with other rodent methods and cannot be explained.

Hydrogen Cyanide

Measurements of the incapacitating effects of HCN with this test method have not been reported.

Rotarod Performance

Various forms of the rotarod have been used for assessing sensorimotor deficits in rodents. The system tests the coordination and endurance of an experimental animal walking on a slowly rotating rod suspended above a grid floor, electrified to motivate the animal to remount the rod if a fall occurs. The rotarod apparatus generally consists of several separate chambers which enclose a 1- to 3-inch diameter rod covered with an anti-skid material. The rod is rotated by an external motor at 4 to 6 rpm. Test animals, trained to remain balanced on the rotating rod, invariably remount the rod within a few seconds if a fall occurs. During exposure tests, an animal is considered incapacitated when, after a fall from the rotating rod, it is unable to remount the rod within a prescribed period of time.

Carbon Monoxide

Using this test system, investigators at The University of Michigan conducted a series of experiments in which rats were exposed to CO concentrations ranging from 1070 to 1445 ppm (average concentration) for 30 minutes [21]. From the data, a dose-response curve was established and an EC_{50} value for incapacitation of approximately 1050 ppm was derived. The Ct value of 31,500 ppm-minutes for this exposure is reasonably close to the range of values estimated for escape impairment in humans and to Ct products obtained with most rodent methods.

The rotarod was also used at Southwest Research Institute to measure the incapacitating effects of CO [12]. Rats were exposed to mean concentrations of 1947 and 3000 ppm CO and times-to-incapacitation were recorded. Mean times-to-incapacitation were 22.4 minutes at 1947 ppm and 12 minutes at 3000 ppm CO, which equate to Ct values of approximately 44,000 and 36,000 ppm-minutes, respectively. These values are within the range of Ct products estimated for human escape impairment, but are somewhat higher than those calculated from the Michigan data.

Hydrogen Cyanide

Measurements of the incapacitating effects of HCN with the rotarod have not been reported.

Shuttlebox

The shuttlebox has been used both at the University of Michigan and Southwest Research Institute in studies on the incapacitating effects of CO [11,21]. The shuttlebox generally consists of a box partitioned into two compartments with a doorway between the compartments. The floor of the box consists of metal rods so as to provide a shock in either compartment. A conditioned stimulus, provided by a light and either a tone or buzzer, is activated for a few seconds. Following the stimulus, the floor grid is then electrified. If the animal moves to the adjacent compartment before the shock is presented, the response is recorded as an avoidance. If the animal moves into the adjacent compartment only after administration of the shock, the response is recorded as escape.

Carbon Monoxide

Using this test system, investigators at the University of Michigan exposed rats to either a low dose (1230–1360 ppm), a middle dose (1410–1840 ppm) or a high dose (2025–2610 ppm) of CO for 30 minutes [21]. At the low dose, the animals did not exhibit impaired avoidance performance until after 25 minutes of exposure; at the high dose, 10 minutes elapsed before avoidance behavior decayed. Loss of escape performance was observed after 15 minutes of exposure to the high dose of CO, with a 50% decrease in the number of escape trials after approximately 29 minutes of exposure. The Ct values for this 50% decrease in escape performance after 29 minutes range from 41,000 to 53,000 ppm-minutes and are comparable to the estimated Ct range for human escape impairment by CO.

The shuttlebox was also used at the Southwest Research Institute in studies on the effects of CO on escape impairment of rats in a paradigm comparable to that used for nonhuman primates [11]. However, because of the lengthy training time required for rats to attain a satisfactory performance level for discrimination, this part of the task was eliminated as being too complex. The escape performance test was initiated by a tone and lights over two levers, followed by a 20-second shock presented 10 seconds after the stimuli. Pressing of either lever opened a door through which the animal could escape into the adjacent chamber. Prior to the start of each escape trial, animals were exposed for 5 minutes to CO at concentrations that ranged from 6090 to 7480 ppm. The percentage of animals that successfully escaped was plotted as a function of CO concentration (Figure 2). From these data, an EC_{50} value of 6780 ppm for escape impairment was determined. This corresponds to a Ct product of 33,900 ppm-minutes, which is remarkably similar to the 34,250 ppm-minutes determined for primates using a comparable shuttlebox avoidance paradigm. These values, although lower than the University of Michigan shuttlebox data, are in agreement with other rodent models and reasonably close to the range estimated for human escape impairment.

Hydrogen Cyanide

The shuttlebox has not been used to study the incapacitating effects of HCN in either rodents or primates.

CONTINUOUS SHOCK-AVOIDANCE LEVER PRESSING

A continuous shock-avoidance lever press method was developed at the Johns Hopkins University to measure behavioral alterations in rats exposed to combustion gases [25]. The apparatus consisted of a plexiglas avoidance chamber equipped with a stainless grid floor and a lever connected to a microswitch mounted on the end wall. A digital computer connected to the chamber controlled the avoidance schedule and recorded the number of shocks and responses. The avoidance schedule provided for the delivery of brief electric shocks to the grid floor every 5 seconds unless the test animal pressed the lever. Each lever press postponed the next shock by 30 seconds. The decrease in the rate of lever pressing and increase in the number of shocks received by an animal were used as measures of behavioral impairment of the animal.

Carbon Monoxide

In experiments conducted with pure CO, rats were exposed to a range of concentrations between 1000 and 2000 ppm for 30-minute periods. Although test animals differed in sensitivity to CO, a significant increase in shock rate was observed at 1300 ppm. The investigators also reported that significant changes in response and shock rates were evident as early as 10–15 minutes at 2000 ppm. These exposures equate to a Ct range of 20,000 to 30,000 ppm-minute, which is somewhat lower than the Ct values for incapacitation of rats obtained with other behavioral test methods. However, it is likely that these performance changes represent a less severe endpoint than the incapacitation endpoints used with other methods.

PHYSIOLOGICAL/OBSERVATIONAL

In addition to the behavioral test methods, physiological and observational techniques have been used in some studies of the incapacitating effects of combustion atmospheres. In the University of Pittsburgh combustion toxicity test, "incapacitation" has been evaluated by measurement of sensory irritation and asphyxiant effects [26]. According to the investigators, the reflex inhibition of respiration resulting from inhalation of a sensory irritant may be used to estimate the incapacitating concentration in humans. Although the respiratory pattern may also be altered by narcotic gases, this endpoint has received limited usage for measurement of incapacitation by pure narcotic gases.

The onset of cardiac arrhythmias was investigated as a measure of incapacitation in rats at the McDonnell Douglas Corporation [22]. In a study in which the animals were exposed to CO, the investigators reported that cardiac arrhythmias occurred slightly prior to incapacitation as determined in an activity wheel.

Observational techniques are used to evaluate the incapacitating effects of combustion atmospheres in the University of San Francisco toxicity test method [27]. Incapacitation is defined as either loss of equilibrium (staggering), prostration, convulsions or collapse. Times-to-incapacitation are recorded on the basis of observations of the first appearance of any of these toxic signs.

COMPARATIVE STUDIES OF TEST METHODS

A comparison between the leg-flexion shock avoidance and rotating activity wheel methods was conducted at the University of Utah using CO [18]. The Ct curves obtained, shown in Figure 5, demonstrate little difference between the two methods. Both result in incapacitating doses of CO in the same Ct range, with the activity wheel perhaps being slightly more sensitive at low concentrations. Blood carboxyhemoglobin saturation at incapacitation for the activity wheel averaged about 50 percent, compared to about 60 percent for the leg-flexion method.

A comparative evaluation of three test methods, tumble cage, rotarod and shuttlebox, was made at the University of Michigan [21]. With the tumble cage and the rotarod, concentration-response curves

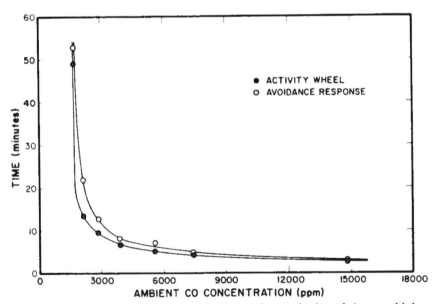

Figure 5. Effect of CO concentration on time to incapacitation of the rat with leg-flexion avoidance and rotating activity wheel methods [18].

were established for incapacitation in rats exposed to CO for 30 minutes. The EC_{50} values demonstrated little difference between the two methods. With the shuttlebox, a time-response curve, based on the percentage of trials in which there was no response to shock, was established. However, this technique does not allow a direct comparison of sensitivity with other methods. The investigators concluded that the tumble cage and the rotarod were essentially equivalent in measuring incapacitation of rats by CO, although the tumble cage was slightly more sensitive, due possibly to the exercising of the test animals. With respect to the shuttlebox, the investigators stated that the shuttlebox offers considerable potential in measuring decrements in performance but that the lengthy (3 to 14 days) training period for rats precludes its use for routine testing.

In a comparison of the rotarod and leg-flexion shock avoidance methods at SwRI, test animals exposed to 1947 ppm CO were incapacitated almost 20% sooner by the rotarod than the leg-flexion method [12]. The Ct values for this CO concentration were approximately 44,000 ppm-min (rotarod) and 59,000 ppm-min (leg-flexion).

CONCLUSIONS

It would appear that most of the described methods for assessment of the incapacitating effects of narcotic fire gases on rats are sensitive to approximately the same ranges of accumulated concentration-time doses and are reasonably close to the ranges deemed potentially hazardous to humans (See Table 3). Experimental rodent data demonstrate that incapacitating doses are about one-third to one-half of those required to effect death, both for CO and for HCN. Choice of a laboratory method should depend on a combination of endpoint objectivity, minimal animal training, ease of execution and cost of experimentation. In view of these criteria, the authors favor the leg-flexion shock avoidance method.

Measurement of the incapacitating effects of the narcotic fire gases is a relevant endpoint for assessing the toxicological hazard of a combustion atmosphere. These effects are undoubtedly responsible for delay or impairment of escape and subsequent death of numerous fire victims. Besides having relevance, incapacitation has a significant advantage over lethality as an endpoint in laboratory test methods with rodents. Incapacitating concentrations in rodents can readily be verified with nonhuman primate studies whereas it is unlikely that lethal concentrations will be similarly verified.

The concentration-time-response relationships for the narcosis-producing toxicants, CO and HCN, are becoming quite well understood and may provide a strategy for predicting toxic effects from analytical studies. Effects of these individual toxicants on rats, along with reasonable extrapolation to humans via nonhuman primate

Table 3. Accumulated doses (Ct products) of narcotic gases
associated with incapacitation*.

Subject/Method	CO Ct ppm-min	HCN Ct ppm-min
Humans	35,000–45,000 (light activity)	750–2500
Baboons (SwRI)	34,000 (active)	
Monkeys (Huntingdon)	25,000 (active) 50,000 (restrained)	1176–1900
Rats**		
Leg Flexion (Utah)	30,000–40,000	
(SwRI)	30,000–40,000	1200–2700
Activity Wheel (FAA)	37,000	800
(Michigan)	30,750	
(McDonnell Douglas)	22,000–36,000	
Pole-Climb Avoidance (SRI)	48,000	
Rotarod (Michigan)	31,500	
(SwRI)	36,000–44,000	
Shuttlebox (Michigan)	41,000–53,000	
(SwRI)	33,900	

*Ct products are presented as single values, rather than ranges, when data for only one concentration or time were available or an EC_{50} was reported.
**Animals are active in all methods except leg-flexion which requires animals to be restrained

studies, can be predicted from analytical determination of the time course of their evolution and consultation and the relevant Ct curves. However, combined effects for CO and HCN cannot yet be predicted with certainty, although current research is aimed at answering questions of any additivity or synergism. In addition, concentration-time-response relationships for irritant combustion products (HCl, in particular) have not been sufficiently worked out to enable prediction of effects from analytical data. Research in this area also is in progress.

In the case of materials which produce toxicants having other mechanisms of action (e.g., neurotoxins), animal exposures will be required in addition to analytical data. However, a simple elementary screening test would be satisfactory, with the number of animals used quite minimal compared to that required for full dose-response testing.

REFERENCES

1. Boettner, E. A., Ball, G. L., and Weiss, B., "Combustion Products from the Incineration of Plastics," Final Report on U.S. Environmental Protection Agency Research Grant No. EDC-00386, The University of Michigan (February 1973).

2. Newman Dorland, W. A., "The American Illustrated Medical Dictionary," 21st Edition, Copyright W. B. Saunders Company (1947).
3. Goodman, L. S., and Gilman, A., *The Pharmacological Basis of Therapeutics*, The MacMillan Company, New York (1965).
4. National Research Council Committee on Fire Toxicology, *Fire Toxicology: Methods for Evaluation of Toxicity of Pyrolysis and Combustion Products*, Report No. 2, Washington, D.C., National Academy of Sciences (1977).
5. Hartzell, G. E., Packham, S. C., and Switzer, W. G., "Assessment of Toxic Hazards of Smoke: Toxicological Potency and Intoxication Rate Thresholds," *Fire and Materials*, Vol. 7, No. 3, pp. 128–131 (1983).
6. Packham, S. C., and Hartzell, G. E., "Fundamentals of Combustion Toxicology in Fire Hazard Assessment," *J. of Testing and Evaluation*, Vol. 9, No. 6, p. 341 (1981).
7. Halpin, B. M., and Berl, W. G., "Human Fatalities From Unwanted Fires," Report NBS-GCR-79-168, National Bureau of Standards, Washington, D.C. (1978).
8. Harland, W. A., and Woolley, W. D., "Fire Fatality Study-University of Glasgow," Information Paper IP 18/79, Building Research Establishment, Glasgow, Scotland (1979).
9. Kimmerle, G., "Aspects and Methodology for the Evaluation of Toxicological Parameters During Fire Exposure," *The Journal of Fire and Flammability Combustion Toxicology Supplement*, Vol. 1 (February 1974).
10. Stewart, R. D., *et al.*, "Experimental Human Exposure to High Concentrations of CO," *Archives of Environmental Health*, Vol. 26, pp. 1–7, (1973).
11. Southwest Research Institute, Unpublished Work.
12. Purser, D. A., "Behavior of Monkeys at High Levels of Carbon Monoxide Related to Effects in Man," *Arch. Env. Hlth.* (In Press).
13. Purser, D. A., and Woolley, W. D., "Biological Studies of Combustion Atmospheres," *J. Fire Sciences*, Vol. 1., No. 2, pp. 118–144 (1983).
14. Purser, D. A., Grimshaw, P., and Berrill, K. R., "Intoxication by Cyanide in Fires: A Study in Monkeys Using Polyacrylonitrile," *Arch. Env. Hlth.* (In Press).
15. Packham, S. C., "Behavior and Physiology: Tools for the Assessment of Relative Toxicity," Paper presented at 17th National Cellular Plastics Conference on Safety and Product Liability, Washington, D.C. (November 18–20, 1974).
16. Product Research Committee, *Fire Research on Cellular Plastics: Status 1979*, Vol. 1, Product Research Committee, Washington, D.C. (1980).
17. U.S. National Bureau of Standards, "Further Development of a Test Method for the Assessment of the Acute Inhalation Toxicity of Combustion Products," NBSIR 82-2532 (June 1982).
18. Hartzell, G. E., Packham, S. C., Hileman, F. D., Israel, S. C., Dickman, M. L., Baldwin, R. C., and Mickelson, R. W., "Physiological and Behavioral Responses to Fire Combustion Products," Fourth International Cellular Plastics Conference, The Society of the Plastics Industry, Incorporated, Montreal, Canada (November 18, 1976).
19. Kaplan, H. L., Grand, A. F., and Hartzell, G. E., *Combustion Toxicology:*

Principles and Test Methods, Technomic Publishing Company, Incorporated, Lancaster, Pennsylvania (1983).

20. Crane, C. R., et al., "Inhalation Toxicology: I. Design of a Small-Animal Test System; II. Determination of the Relative Toxic Hazards of 75 Aircraft Cabin Materials," Report No. FAA-AM-77-9 Prepared for Office of Aviation Medicine, Federal Aviation Administration (March 1977).

21. Boettner, E. A., and Hartung, R., "The Analysis and Toxicity of the Combustion Products of Natural and Synthetic Materials," Final Report for The Society of the Plastics Industry, Incorporated and The Manufacturing Chemists Association (July 31, 1978).

22. Gaume, J. G., Reibold, R. C., and Speith, H. H., "Initial Tests of the Combined ECG/Ti Animal Systems Using Carbon Monoxide Exposure," *Journal of Combustion Toxicology,* Vol. 2, p. 125 (May 1981).

23. Dilley, J. V., Martin, S. B., McKee, R. and Pryor, G. T., "A Smoke Toxicity Methodology," *J. Comb. Tox.,* Vol. 6, pp. 20-29 (1979).

24. SRI International Life Sciences Research Report, "Smoke Hazards Research," Volume 8, Number 3 (July 1978).

25. Zoltan Annau, "Behavioral Toxicity of Cellular Plastics," Final Report for Products Research Committee, National Bureau of Standards, Grant No. RP-77-U-2, by The Johns Hopkins University (June 1979).

26. Alarie, Y. C., and Anderson, R. C., "Toxicologic and Acute Lethal Hazard Evaluation of Thermal Decomposition Products of Synthetic and Natural Polymers," *Toxicology and Applied Pharmacology,* Vol. 51, pp. 341-362 (August 1979).

27. Hilado, C. J., and Schneider, J. E., "Toxicity of Pyrolysis Gases From Polytetrafluoroethylene," *Journal of Combustion Toxicology,* Vol. 6, p. 91 (May 1979).

Evaluation of Smoke Toxicity Using Concentration-Time Products

GEORGE V. ALEXEEFF

Fire Technology Laboratory
Weyerhaeuser Company
P.O. Box 188
Longview, Washington 98632

STEVEN C. PACKHAM

BETR Sciences, Inc.
996 South 1500 East
Salt Lake City, Utah 84105

ABSTRACT

A variety of protocols and test methods have been developed to examine combustion product toxicity of building and furnishing materials. To promote comparisons of results among such tests, a standard method for quantifying exposure intensity as an estimate of dosage was developed. Smoke concentrations (C) were calculated from sample weight-loss data and integrated over the animal exposure time (t), such that Ct products were obtained in units of "mg/L · minutes." Application of the method was demonstrated by calculating Ct products from data reported in the literature for forty-five materials examined by the University of Pittsburg test (UPT) method.

Smoke concentrations from individual materials tested in the UPT method were not constant and some materials produced smoke for only one third of the 30-minute exposure period. Some smoke exposures calculated using the proposed method were as low as one-fourth to one-sixth of that implied by the reported LC_{50} (g) values. Expression of toxic potency in Ct units [$L(Ct)_{50's}$] changed the relative ranking of some materials. Generally similar materials exhibited similar $L(Ct)_{50's}$. The range of $L(Ct)_{50's}$, especially for wood products, was found to be narrow when compared to that of pure gases and vapors.

INTRODUCTION

THIS PAPER EXAMINES the quantification of exposure in tests intended to measure the relative toxicity of smoke produced from burning materials. In an attempt to understand causes for and reduce the number of fire related deaths caused by smoke inhalation poisoning, a variety of protocols and test methods have been developed to examine the toxicity of smoke produced from home furnishings and building materials (Table 1). The International Standards Organization (ISO) through its Technical Committee 92 has established a special subcommittee (SC-3) to evaluate smoke toxicity methods in preparation for drafting international standards in this area. Also, the American Society for Testing and Materials Subcommittee on Smoke and Combustion Products has established several task groups to develop a consensus standard method. At least ten states have introduced legislation to regulate use of commercial materials on the basis of the toxicity of the smoke they produce when burned (Table 2). Recently, authorities in the State of New York recommended establishing smoke toxicity data banks for materials using the University of Pittsburg test method. For all of these preregulatory processes it is essential that scientifically based prerequisites for protocols be determined. Quantification of exposure lies at the heart of all inhalation toxicity testing and for the combustion toxicologist it poses a formidable challenge. In this paper, a procedure for standardizing exposure intensity is proposed.

Table 1. Methods used to test smoke toxicity[a].

Federal Republic of Germany (DIN) Method
Federal Aviation Administration (FAA) Method
National Bureau of Standards (NBS) Method
Radiant Heat Test Method
University of Pittsburgh Method
University of San Francisco Method
Dow Chemical Company Test Method
Harvard Medical School Test Method
Japanese Combustion Toxicity Test Method
McDonnell-Douglas Corporation Test Method
Stanford Research Institute International Test Method
U.S. Testing Company, Incorporated Test Method
University of Michigan Test Method
University of Tennessee Test Method
University of Utah Test Method
USSR Test Method[b]
Carnegie Mellon Institute of Research (CMIR) Method[c]

[a]All methods described in Kaplan *et al*, 1981 unless otherwise indicated
[b]International Mariners Organization FP/353 19 Jan 84
[c]Gad *et al*. 1983

Table 2. Legislation proposing smoke toxicity testing in the United States.

State	Sponsored By	Bill Number	Date Introduced
California	Davis	AB 1884	03/04/83
	Waters	AB 1535	03/03/83
Maryland	Sher	HB 1274	02/03/84
		HB 1588	
New York	Dunne	S 8988	03/30/82
Massachusetts	White	HB 2783	01/04/84
	Flaherty	HB 4793	01/04/84
	Atkins	SB 2201	01/04/84
		Section 64	
Connecticut	Senate Public Safety Comm.	SB 109	02/10/84
New Jersey	Rooney	AB 1108	02/06/84
Indiana	Mosley	SN 118	01/04/84
Minnesota	Clark/Munger	HB 865	03/24/83
	Diessner	SB 1603	03/06/84
Pennsylvania	Petrone	HB 142	10/12/83
Texas	Hill	HB 1620	02/10/83

BACKGROUND

The development of combustion toxicology as a formal science was preceded historically by an awareness that smoke inhalation, rather than heat, was the principal cause of death for most fire victims. Recent statistics confirm that of all fire fatalities (currently, about 6000 annually in the United States) approximately 80 percent were from smoke inhalation [28]; most deaths were associated with carbon monoxide poisoning.

In their 1977 report [13], the National Academy of Sciences Committee on Fire Toxicology identified Zapp's 1951 publication [29], "The Toxicity of Fire," as marking the beginning of combustion toxicology. It wasn't until the 1960's, however, that a substantial number of other scientists became involved. Interest in regulation was sparked in 1974 when researchers at the University of Utah identified a neuroleptic organophosphate in the smoke of a noncommercial polyurethane foam [26]. In the last ten years there has been a rapid increase in the number of scientists and research institutes specializing in combustion toxicity.

In his 1968 paper, MacFarland [22] identified the quantification of dose as one of the most technically difficult aspects of combustion toxicology, the principal complication being the nonhomogeneous nature of smoke. Burning materials can yield gases and aerosols, the relative proportions of which change as a function of temperature and mode of combustion. A second complicating factor, shared by other fields of inhalation toxicology, is that the amount of toxicant absorbed changes with each breath. MacFarland [23] and others [14,25] have suggested that while the true dose cannot be measured directly, it is reasonable to assume that the accumulated body burden of toxicants from smoke inhalation is a function of smoke concentration (C) and exposure time (t). Smoke concentration has been traditionally expressed as the mass of material lost from the sample per chamber volume [19,25]. Exposure time has usually been preset to 30 minutes [1,15,18,19]. Since both parameters can be controlled or estimated, the magnitude of the exposure, or "exposure dose," can be expressed as the "Ct product," which is mathematically defined as $\int C(t)dt$. This can be used in place of "true dose" to develop dose-response curves and as input data for traditional statistical analysis.

Toxicity values based on "concentration" and "time" quantifications have been reported by Hilado and Huttlinger [18] and Alarie and Anderson [2]. However, the Ct products described by these investigators were not based on precisely the same variables proposed by MacFarland [23] as they defined "time" as the time to an observed end-point rather than actual exposure time. Exposure time may be equivalent to "time-to-effect" only if the response occurs before the end of the exposure; this may not be the case, particularly when considering post-exposure mortality. Alarie and Anderson [2] defined concentration in terms of weight of material placed in the combustion furnace (g) rather than mass of material distributed per volume of air (mg/L).

Neither the Ct product nor any other exposure quantification technique has been standardized in combustion toxicology. Rank comparison of test method results conducted by Anderson *et al.* [6,7,8] indicated only partial agreement between the UPT and the National Bureau of Standards Cup method. Even though both methods report data based on a 30-minute exposure, the absence of agreement in LC_{50}'s may have resulted from different effective animal exposure conditions caused by their respective methods for heating the test specimen and their different smoke handling techniques. In the authors' opinion, the first step in comparing test results from two or more test methods would be the conversion of exposure parameters to similar units (e.g. mg/L · minutes) so that equivalencies in exposure intensity or exposure dose can be determined. Results as currently reported by test methods listed in Table 1 appear in a variety of formats. Exposure intensities are therefore difficult, if not impossible, to compare. Further, several methods rank materials on the basis of mass loading and it is not clear how closely this followed exposure intensity. If exposure intensity could be calculated for several test methods by a standard procedure, the science of combustion toxicology would benefit from valuable interlaboratory comparisons of test method results. The principle demonstrated in this paper for converting smoke toxicity data to Ct products could effectively be applied to other test methods in addition to the UPT [e.g. 4,15] using smoke concentration estimates.

METHODS

Ct products were calculated for forty-five materials from data previously developed using the University of Pittsburg test (UPT) method [1,2,3,6,7,21]. The reasons for choosing the University of Pittsburg test method were straightforward. First, the procedure's continuous weight loss measurements allowed estimation of smoke concentrations. Second, the results have been reported in great detail in the literature making such calculations possible. Third, the published data constitute a substantial number and range of materials. And finally, it has been recommended that materials be tested using this method as a requirement for marketing in New York State.

In the UPT method, a quantity of test sample is placed in a Lindberg furnace programmed to increase its temperature 20 °C per minute. Pyrolysis and combustion products are diluted into a 20L per minute airflow which passes through a 2.3L animal exposure chamber under dynamic conditions. Although information about percent of weight remaining versus temperature was reported (Figure 1), developers of the test expressed LC_{50}, in terms of initial sample weight (g).

Calculations of exposure, as described by MacFarland [23], produce Ct products in units such as "mg/L · minutes." Since both heating rate (20 °C/min) and air flow (20L/min) were related to a common time denominator in the UPT method, the percent weight remaining versus temperature curves (Figure 1) were converted to "concentration" versus time curves (Figure 2). Each curve was divided into six consecutive 5-minute time periods. The areas under the concentration-time histograms were then determined to yield the Ct products expressed in mg/L · minutes. Thus, the integral of the concentration-time curve was approximated. The percent residue remaining for each material was obtained from the curves or from reported tables.

RESULTS AND DISCUSSION

Data obtained from published UPT results were analyzed to determine the percent of initial sample weight which remained following a typical 30-minute test (Table 3). Residual sample weights varied from 0 percent for PTFE resin, urea formaldehyde, and cellular plastics (identified as GM-21, GM-43, GM-47, and GM-49) to over 85 percent for THHN coated wire. These data suggest that on the basis of the amount of test specimen that was converted into smoke, the exposure

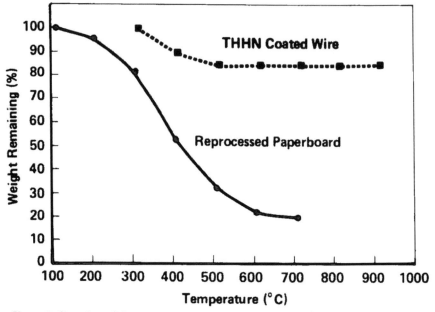

Figure 1. Sample weight versus temperature curves plotted for two materials tested near their LC_{50} using the University of Pittsburg test method for assessing the toxicity of combustion products of materials. These curves simulate those originally plotted by Anderson et al. (1983c) where the $LC_{50's}$ of reprocessed paperboard and THHN coated wire were reported to be 22.1 and 22.0 grams respectively.

could have been as low as one-sixth of that expected from the reported LC_{50} (g) value alone. Thus, materials which left a substantial residue, produced combustion products that were relatively more toxic than would be inferred from the $LC_{50's}$ (g) which described the quantity of material placed into the furnace.

Exposure profiles, produced by individual materials, varied substantially over the six consecutive 5-minute time windows of the test (Table 4). Some materials lost no weight during extended portions of the experiment. For example, four materials (THHN coated wire, PTFE resin, GM-47, and GM-49) only lost weight during 10 minutes of the test. Thus, within this data a three-fold difference was discovered in meaningful exposure time, since the technique used a flow through system. The variability in exposure time and in expected smoke concentration (documented above), could lead to a potential eighteen-fold difference in the calculated toxic potency of the smoke.

None of the materials produced a constant concentration during the 30-minute exposure. In fact, thirty-eight of the forty-five materials lost more than 50 percent of their weight within a 10-minute segment

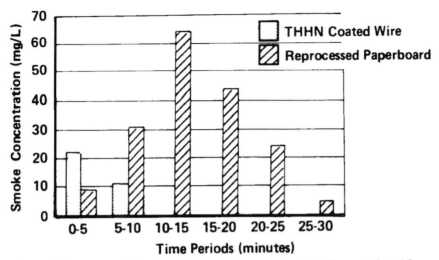

Figure 2. A representation of the calculated smoke concentrations produced from reprocessed paperboard and THHN coated wire during the tests. These were calculated using the data reported by Anderson et al., (1983c) as shown in Figure 1. Each experiment was divided into six 5-minute periods for which the average smoke concentration, based on sample weight loss, was determined and plotted. The LC_{50}s reported for reprocessed paperboard and THHN coated wire were 22.1 and 22.0 grams respectively. However, the calculated smoke exposures were 883.5 and 165.0 (mg/L) · minutes respectively.

Table 3. Sample residue following LC_{50} exposure using the University of Pittsburg Test Method.

Material	Percent Weight Remaining (Mean ± SD)	Source
PTFE Resin	0	Barrow et al., 1979
Urea Formaldehyde	0	Alarie and Anderson, 1979
GM-21	0	Ibid.
GM-43	0	Ibid.
GM-47	0	Ibid.
GM-49	0	Ibid.
ABS Pipe	0.3 ± 0.3	Anderson et al., 1983c
Polyurethane Foam	0.5 ± 0.4	Ibid.
GM-23	1	Alarie and Anderson, 1979
GM-35	1	Ibid.
GM-25	3	Ibid.
PVC-A	5	Barrow et al., 1979
Polyurethane	5	Alarie et al., 1983
Fyreprufe™ Ponderosa	5	Lieu et al., 1982
High Density Fiber	6.1 ± 4.1	Anderson et al., 1983c
GM-37	8	Alarie and Anderson, 1979

(continued)

Table 3. (continued)

Material	Percent Weight Remaining (Mean ± SD)	Source
Cotton	8	Alarie et al., 1983
Polyester	9	Ibid.
GM-27	10	Alarie and Anderson, 1979
GM-29	10	Ibid.
GM-31	11	Ibid.
CPVC Pipe	13.3 ± 5.9	Anderson et al., 1983c
GM-41	14	Alarie and Anderson, 1979
Cotton Fabric	14.4 ± 5.6	Anderson et al., 1983c
Wood Base Ceiling Tile	15.5 ± 3.2	Ibid.
Asphalt Felt	16.3 ± 3.0	Ibid.
Reprocessed Paper Board	17.5 ± 3.5	Ibid.
NCX Douglas Fir	18	Alarie et al., 1983
Douglas Fir	19.3 ± 1.1	Anderson et al., 1983c
Douglas Fir	20	Alarie and Anderson, 1979
Cellulose Fiber	21	Ibid.
Paper Wall Cover	21.1 ± 1.7	Anderson et al., 1983c
Wool	22.0	Alarie and Anderson, 1979
Vinyl Wall Cover	24.7 ± 2.0	Anderson et al., 1983c
PVC Conduit	29.4 ± 2.6	Ibid.
Nylon Carpet-Jute	33.3 ± 2.1	Ibid.
PTFE Wire	38.9 ± 2.0	Ibid.
Intumescent Paint	42.6 ± 0.6	Ibid.
Phenol Formaldehyde	52.0	Alarie and Anderson, 1979
Nylon Carpet-Foam	57.9 ± 3.3	Anderson et al., 1983c
Vinyl Coated Wire	68.2 ± 0.4	Ibid.
Mineral Base Ceiling Tile	74.2 ± 0.4	Ibid.
Latex Paint	75.3 ± 0.6	Ibid.
Gypsum Wall Board	76.1 ± 0.2	Ibid.
THHN Coated Wire	85.1 ± 0.6	Ibid.

*The GM designations are as follows GM-21, flexible polyurethane foam; GM-23, same as GM-21 with fire retardant; GM-25, high resilience flexible polyurethane foam; GM-27, same as GM-25 with fire retardant; GM-29, rigid polyurethane foam; GM-31, same as GM-29 with fire retardant; GM-36, fluorocarbon blown rigid polyurethane foam; GM-37, CO_2 blown rigid polyurethane foam; GM-41, rigid isocyanurate foam; GM-43, GM-41 with some polyurethane; GM-47, expanded polystyrene; GM-49, same as GM-47 with fire retardant

Table 4. Average smoke concentrations (mg/L) calculated during five-minute time intervals for materials tested using the University of Pittsburg Test Method.

Material	0-5	5-10	10-15	15-20	20-25	25-30	Source
PTFE Resin	0	5.4	1.0	0	0	0	Alarie & Anderson, 1979
PTFE Coated Wire	5.6	4.5	1.8	0.6	1.5	2.1	Anderson et al., 1983c
Wool	2.4	3.6	4.8	5.1	4.5	3.0	Alarie & Anderson, 1979
C-PVC Pipe	12.2	6.2	2.4	1.4	1.6	1.6	Anderson et al., 1983c

(continued)

Table 4. (continued)

Material	0-5	5-10	10-15	15-20	20-25	25-30	Source
Phenol Formaldehyde	1.9	2.5	4.4	8.2	6.9	6.3	Alarie & Anderson, 1979
Urea Formaldehyde	1.3	4.8	9.3	5.8	2.8	1.3	Ibid.
THHN Coated Wire	22.0	11.0	0	0	0	0	Anderson et al., 1983c
Vinyl Coated Wire	16.8	9.9	2.0	1.0	2.0	2.0	Ibid.
PVC Conduit	15.9	11.8	5.9	4.1	4.1	0.6	Ibid.
Intumescent Paint	3.3	9.0	14.8	11.5	5.7	1.6	Ibid.
Paper Wall Cover	4.6	11.2	21.8	9.2	4.6	0.7	Ibid.
GM-41*	3.2	14.7	12.2	10.2	14.7	0	Alarie & Anderson, 1979
ABS Pipe	1.1	11.8	24.6	9.0	6.7	2.8	Anderson et al., 1983c
GM-47	0	5.2	52.8	0	0	0	Alarie & Anderson, 1979
Nylon Carpet-Foam	22.7	14.2	7.1	2.1	5.0	8.5	Anderson et al., 1983c
GM-43	11.0	12.2	6.7	14.6	11.6	4.9	Alarie & Anderson, 1979
Vinyl Wall Cover	2.1	17.0	29.3	7.7	5.1	1.3	Anderson et al., 1983c
GM-31	18.0	24.6	9.8	10.7	9.0	0	Alarie & Anderson, 1979
GM-37	9.6	34.4	20.8	4.0	3.2	1.6	Ibid.
GM-35	16.5	27.0	12.8	9.8	6.8	1.5	Ibid.
Mineral Base Ceiling Tile	8.9	17.7	35.4	8.9	5.9	0	Anderson et al., 1983c
GM-25	4.2	19.1	49.8	3.3	3.3	0	Alarie & Anderson, 1979
Latex Paint	6.0	11.9	26.8	20.9	11.9	3.0	Anderson et al., 1983c
High Density Fiberboard	16.2	22.5	36.0	4.5	3.6	0	Ibid.
Polyurethane Foam	7.7	14.5	23.0	23.0	13.6	3.4	Ibid.
GM-29	21.8	32.4	14.6	12.5	9.4	0	Alarie & Anderson, 1979
Cellulose Fiber	9.5	47.6	16.7	9.5	7.1	2.4	Ibid.
Nylon Carpet-Jute	5.6	5.6	9.9	42.3	25.4	4.2	Anderson et al., 1983c
Asphalt Felt	2.4	19.2	48.0	16.8	7.2	3.6	Ibid.
Wood Base Ceiling Tile	12.4	28.3	41.8	10.2	4.5	0	Ibid.
GM-49	0	2.0	98.0	0	0	0	Alarie & Anderson, 1979
GM-23	9.4	4.2	79.0	6.2	3.1	0	Ibid.
Cotton Fabric	24.3	31.1	35.1	9.5	8.1	4.1	Anderson et al., 1983c
Polyurethane	1.3	21.0	41.9	51.1	7.9	1.3	Alarie et al., 1983
GM-27	14.4	14.4	92.2	4.3	2.9	0	Alarie & Anderson, 1979
GM-21	15.5	12.9	90.3	7.7	2.6	0	Ibid.
PVC-A	68.4	19.8	25.8	15.2	10.6	3.3	Ibid.
Reprocessed Paperboard	8.8	30.9	64.1	44.2	24.3	4.4	Anderson et al., 1983c
Fyreprufe™ Ponderosa Pine	11.2	66.9	129.3	2.2	2.2	0	Lieu et al., 1982
NCX Douglas Fir	16.8	67.2	67.2	33.6	33.6	33.6	Ibid.
Polyester	75.6	156.8	14.0	5.6	2.8	0	Alarie et al., 1983
Cotton	70.8	113.2	39.6	17.0	14.2	5.7	Ibid.
Douglas Fir	15.5	46.5	77.5	62.0	52.7	34.1	Anderson et al., 1983c
Gypsum Wallboard	63.0	88.2	126.0	12.6	12.6	0	Ibid.
Compressed Spruce-Pine	24.4	73.1	121.8	97.4	82.8	53.6	Alarie & Anderson, 1981
Douglas Fir	12.8	19.1	82.9	172.3	178.6	63.8	Alarie & Anderson et al., 1979

*The GM designations are as follows: GM-21, flexible polyurethane foam; GM-23, same as GM-21 with fire retardant; GM-25, high resilience flexible polyurethane foam; GM-27, same as GM-25 with fire retardant; GM-29, rigid polyurethane foam; GM-31, same as GM-29 with fire retardant; GM-35, fluorocarbon blown rigid polyurethane foam; GM-37, CO_2 blown rigid polyurethane foam; GM-41, rigid isocyanurate foam; GM-43, GM-41 with some polyurethane; GM-47, expanded polystyrene; GM-49, same as GM-47 with fire retardant.

210

of the test, and eighteen of them (PTFE, GM-47, GM-37, GM-25, GM-49, GM-23, polyurethane, GM-27, GM-21, Fyreprufe™ ponderosa pine, polyester, cotton, vinyl coated wire, vinyl wall cover, high density fiberboard, wood ceiling tile, nylon carpet with jute backing, and gypsum wallboard) lost more than 70 percent of their weight during that time. Maximum smoke concentration occurred at various stages of the 30-minute period for different materials (Table 4). For example, in Figure 2 the smoke concentration produced from THHN coated wire peaked within 5 minutes while that of reprocessed paperboard peaked at the 10 to 15 minute time period. The importance of the peak locations is not known. However, it is conceivable that early production of an irritant could influence respiration rate and subsequent toxicant uptake.

The lethal concentration-time products $L(Ct)_{50's}$ calculated for the materials are presented in Table 5 along with their reported $LC_{50's}$ (g). The rank order of relative smoke toxicity expressed as $L(Ct)_{50}$ (mg/L · minutes) differs from the LC_{50} (g) values. The THHN and the vinyl coated wires produced very similar $L(Ct)_{50's}$ (165.0 and 168.5 mg/L · minutes respectively) but the $LC_{50's}$ based on the initial sample weight were substantially different (22.0 and 9.9 grams respectively). Douglas fir and gypsum wallboard also had similar effective $L(Ct)_{50's}$ exposures (1441.5 and 1512.0 mg/L · minutes respectively) but very different LC_{50} values (31.0 and 126.0 grams respectively). Conversely, some materials with reportedly similar LC_{50} (g) values produced dramatically different exposure profiles and lethal Ct products. The nylon carpets with foam and jute backing had produced almost identical $LC_{50's}$ (14.2 and 14.1 grams respectively) but their $LC_{50's}$ differed by over 40 percent (298.0 and 426.0 mg/L · minutes respectively). A similar difference was seen between THHN coated wire and reprocessed paperboard. Their $LC_{50's}$ (g) were almost identical (22.0 and 22.1 grams, respectively) but their calculated $L(Ct)_{50's}$ differed by over 400 percent (165.0 and 883.5 mg/L respectively; see Figure 2 also). Other materials had LC_{50} (g) and $L(Ct)_{50}$ values which were similar in both evaluations. For example, mineral based ceiling tile and latex paint had similar lethal Ct products (384.0 and 402.5 mg/L-minutes respectively) and similar LC_{50} values (29.5 and 29.8 grams respectively).

While the magnitude of the exposure can be expressed as a function of C and t, caution must be exercised when comparing $L(Ct)_{50's}$. A review of CO and HCN LC_{50} data [20] indicated that $LC_{50's}$ determined for these gases over a rather limited range of exposure times produced different $L(Ct)_{50's}$. Presumably, this would also be true for smoke. In the case of CO and HCN, the Ct products for a 10-minute LC_{50} and a 30-minute LC_{50} differed by a factor of two. Thus, two materials expressing identical $L(Ct)_{50's}$ when evaluated using one exposure time period may not have similar $L(Ct)_{50's}$ when tested using a different exposure duration. For this reason, $L(Ct)_{50}$ values associated with only one of many possible combinations of exposure parameters, such as those listed in Table 5, should not be viewed as an absolute rank order of toxic potency.

In Table 6, lethal Ct products have been compiled from the literature for thirty-six pure gases/vapors in tests characterized by constant exposure concentrations and equivalent exposure times (30 minutes). Results were tabulated in units as originally reported and then transformed to mg/L · minutes units to facilitate a direct comparison with Ct products developed from the UPT data. The range of toxicity was from 4.5 mg/L · minutes for HCN to 16,500 mg/L · minutes for 1-bromohexane representing a scale traversing 3.6 orders of magnitude. By contrast, the range of toxicity for smoke (Table 5) was from 32.0 mg/L · minutes for PTFE Resin to 2,647.5 mg/L · minutes for Douglas fir representing maximum differences of 1.9 orders of magnitude. The range of toxicity for wood products was less than one order of magnitude. The ranges were compared graphically in Figure 3.

Table 5. Lethal exposure values [L(Ct)$_{50}$s] and their corresponding LC$_{50}$s (g) values as reported in the literature.

Material	L(Ct)$_{50}$ [(mg/L · minute)]	LC$_{50}$ (g)	Source
PTFE Resin	32.0	0.6	Alarie & Anderson, 1979
PTFE Coated Wire	85.0	3.0	Anderson et al., 1983b
Wool	117.0	3.0	Alarie & Anderson, 1979
C-PVC Pipe	127.0	2.7	Anderson et al., 1983b
Phenol Formaldehyde	151.0	6.3	Alarie & Anderson, 1979
Urea Formaldehyde	155.5	2.5	Ibid.
THHN Coated Wire	165.0	22.0	Anderson et al., 1983b
Vinyl Coated Wire	168.5	9.9	Ibid.
PVC Conduit	212.0	5.9	Ibid.
Intumescent Paint	229.5	8.2	Ibid.
Paper Wall Cover	260.5	6.6	Ibid.
GM-41*	275.0	6.4	Alarie & Anderson, 1979
ABS Pipe	280.0	5.6	Anderson et al., 1983b
GM-47	290.0	5.8	Alarie & Anderson, 1979
Nylon Carpet-Foam	298.0	14.2	Anderson et al., 1983b
GM-43	305.0	6.1	Alarie & Anderson, 1979
Vinyl Wall Cover	312.5	8.5	Anderson et al., 1983b
GM-31	360.5	8.2	Alarie & Anderson, 1979
GM-37	368.0	8.0	Ibid.
GM-35	372.0	7.5	Ibid.
Mineral Base Ceiling Tile	384.0	29.5	Anderson et al., 1983b
GM-25	398.5	8.3	Alarie & Anderson, 1979
Latex Paint	402.5	29.8	Anderson et al., 1983b
High Density Fiberboard	414.0	10.5	Ibid.
Polyurethane Foam	426.0	8.5	Ibid.

(continued)

Table 5. (continued)

Material	L (Ct)$_{50}$ [(mg/L · minute)]	LC$_{50}$ (g)	Source
GM-29	453.5	10.4	Alarie & Anderson, 1979
Cellulose Fiber	464.0	11.9	Ibid.
Nylon Carpet-Jute	465.0	14.1	Anderson et al., 1983b
Asphalt Felt	486.0	12.0	Ibid.
Wood Base Ceiling Tile	486.0	11.3	Ibid.
GM-49	500.0	10.0	Alarie & Anderson, 1979
GM-23	509.5	10.4	Ibid.
Cotton Fabric	561.0	13.5	Anderson et al., 1983b
Polyurethane	662.5	13.5	Alarie et al., 1983b
GM-27	641.0	14.4	Alarie & Anderson, 1979
GM-21	645.0	12.9	Ibid.
PVC-A	715.0	15.2	Ibid.
Reprocessed Paperboard	883.5	22.1	Anderson et al., 1983b
Fyreprufe™ Ponderosa Pine	1059.0	22.3	Lieu et al., 1982
NCX Douglas Fir	1110.0	33.6	Ibid.
Polyester	1274.0	28.3	Alarie et al., 1983b
Cotton	1302.0	28.0	Ibid.
Douglas Fir	1441.6	31.0	Anderson et al., 1983b
Gypsum Wallboard	1521.0	126.0	Ibid.
Compressed Spruce-Pine Fir	2265.5	48.7	Alarie & Anderson, 1981
Douglas Fir	2647.5	63.8	Alarie & Anderson et al., 1979

*The GM designations are as follows: GM-21, flexible polyurethane foam; GM-23, same as GM-21 with fire retardant; GM-25, high resilience flexible polyurethane foam; GM-27, same as GM-25 with fire retardant; GM-29, rigid polyurethane foam; GM-31, same as GM-29 with fire retardant; GM-35, fluorocarbon blown rigid polyurethane foam; GM-37, CO_2 blown rigid polyurethane foam; GM-41, rigid isocyanurate foam; GM-43, GM-41 with some polyurethane; GM-47, expanded polystyrene; GM-49, same as GM-47 with fire retardant.

General practice in toxicology has been to classify materials on the basis of LC$_{50}$ or LD$_{50}$ values that differ by an order of magnitude or a factor of 10. Anderson and Alarie [5] stated that while factors below 3 or 4 may be statistically significant they have "no practical importance" when considering the rating of materials. In this analysis, smoke from all solid woods tested were found to differ by only a factor of 2.5 while smoke from processed woods differed by only a factor of 3.4. Consequently, wood products (and possibly some specific generic classes of synthetics) exhibit ranges of toxic potency within which practical differences in toxicity are indistinguishable when evaluated by that standard.

SUMMARY AND CONCLUSIONS

A method has been proposed for calculating exposure intensity in combustion toxicology on the basis of quantitative rate of sample weight loss and time. The method was demonstrated by its application to published test results from the UPT method. Results showed that differences between exposures inferred from starting sample weights and calculated exposures could be greater than an order of magnitude. When toxic potency was expressed as the Ct product based on weight-loss rates, the relative ranking of some materials changed dramatically. It was further observed that materials having similar chemical characteristics tended to yield similar Ct products. A further benefit of the described method was that the smoke exposures could then be directly compared with those from pure gases/vapors. The range of Ct products from smoke was found to be narrow when compared to the range of Ct products obtained from pure gases/vapors. Although the procedure is applicable to any test method which records the rate of sample weight loss, this study was limited to the UPT method to demonstrate the principle. A number of the findings from this analysis relate to specific characteristics of the UPT method, such as its flow through system. However, other toxicity test systems may produce similar results or exhibit limitations not common to the UPT method. In the authors' opinion, the greatest limitation by a combustion toxicity test system

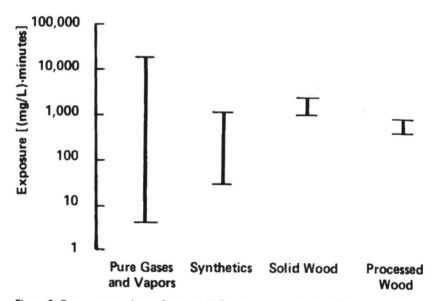

Figure 3. Range comparison of reported LC_{50}'s in exposure units of (mg/L) · minutes for pure gases and for materials tested in the University of Pittsburg test method for assessing the toxicity of combustion products. The pure gases data reflect the range of LC_{50}'s reported for the rats exposed for thirty minutes. The data are listed in Table 6. The LC_{50}'s for the synthetics, solid wood and processed wood, were developed by the University of Pittsburg Test and are listed in Table 3.

Table 6. Reported LC_{50}s and calculated $L(Ct)_{50}$s in the rat for 30-minute exposure periods.

Material	LC_{50} (as reported)	$L(Ct)_{50}$ Exposure [(mg/L) · minutes]	Source
Hydrogen Cyanide	4530 mg/m³ · min	4.5	Ballantyne, 1984
Cobalt Hydrocarbonyl	165 mg/m³	5.0	NIOSH, 1984
Ketene	130 ppm	6.7	Clayton and Clayton, 1981a
Nickel Carbonyl	35 ppm	7.3	NIOSH, 1984
Acrolein	0.3 mg/L	9.0	Skog, 1950
Fluorine	420 mg/m³	12.6	Clayton and Clayton, 1981b
Nitrogen Dioxide	250 ppm	14.1	Hilado and Cummings, 1977
Nitric Acid (white fuming)	244 ppm	18.9	NIOSH, 1984
Trichloroacryloyl Chloride	107 ppm	25.5	NIOSH, 1984
Formaldehyde	1 mg/L	30.0	Skog, 1950
Hydrogen Fluoride	2042 ppm	50.1	Hilado and Cumming, 1977
(E)-Crotonaldehyde	4000 mg/m³	120.0	NIOSH, 1984
Hexachloropropene	425 ppm	129.7	NIOSH, 1984
Carbon Monoxide	155520 mg/m³ · min	155.5	Ballantyne, 1984
Hexafluoropropanone	900 ppm	183.3	Gillies and Rickard, 1984
Carbon Monoxide	5550 ppm	189.6	Hilado and Cumming, 1977
Hydrogen Chloride (vapor)	4701 ppm	210.3	NIOSH, 1984
Hydrogen Chloride (aerosol)	5666 ppm	253.5	NIOSH, 1984

(continued)

Table 6. (continued)

Material	LC$_{50}$ (as reported)	L(Ct)$_{50}$ Exposure [(mg/L) · minutes]	Source
3-Bromopropene	10000 mg/m³	300.0	NIOSH, 1984
Methyl Bromide	11 mg/L	330.0	Bakhishev, 1973
3-Chloro-2-Methylpropene	34000 mg/m³	1020.0	NIOSH, 1984
Acetaldehyde	37 mg/L	1100.0	Skog, 1950
Propionaldehyde	62 mg/L	1860.0	Skog, 1950
Iodoethane	65000 mg/m³	1950.0	NIOSH, 1984
1-Iodopropane	73000 mg/m³	2190.0	NIOSH, 1984
Dichloromethane	88000 mg/m³	2640.0	NIOSH, 1984
Chloromethane	152000 mg/m³	4560.0	NIOSH, 1984
Butyraldehyde	174 mg/L	5220.0	Skog, 1950
1-Iodo-2-Methylpropane	216000 mg/m³	6480.0	NIOSH, 1984
1-Bromobutane	237000 mg/m³	7110.0	NIOSH, 1984
1-Iodobutane	237000 mg/m³	7110.0	NIOSH, 1984
1-Bromopropane	253000 mg/m³	7590.0	NIOSH, 1984
2-Iodopropane	320000 mg/m³	9600.0	NIOSH, 1984
Dichlorodifluoromethane	80 pph	10788.0	NIOSH, 1984
1, 2-Dichloro-1,1,2,2-Tetrafluoroethane	72 pph	15099.7	NIOSH, 1984
1-Bromohexane	550000 mg/m³	16500.0	NIOSH, 1984

would be its inability to provide an estimate for smoke concentration throughout the test.

The original stated purpose for developing the UPT method (as well as other test methods) was to assess the hazard of the material and not of its smoke [5]. The aim of such a hazard evaluation was to screen for supertoxic materials. However, current trends have shifted from "one test hazard evaluation" to complex computer modeling of fire hazard using empirically established input data on factors including smoke density, rate of heat release, and ignitability along with toxicity. This is evidenced by the extensive work being conducted in computer modeling at the National Bureau of Standards, Ohio State University, and Factory Mutual among others. To be optimally compatible, smoke toxicity data must represent a discrete variable in the integrated fire model. Specifically, smoke toxicity values should not be confounded artificially by combustibility properties of the material, sample size, or sample configuration. It is our conclusion that the Ct product has greater application in integrated hazard models than the relative ranking of materials on the basis of LC_{50} values since confounding variables such as sample residue and rate of mass loss are taken into consideration.

Observations made in this study have further implications for the use of smoke toxicity test results in product regulation and pre-regulatory data filing. Clearly the value of reported results is dependent on their usefulness in product comparisons and in decision making when considering product applications. The simple accumulation of data is not a meaningful end in itself. Consequently, in an effort to further the progress of combustion toxicology, a hard look must be taken to determine the type of data necessary for fire risk-assessment and the ability of a test apparatus to produce that information.

ACKNOWLEDGEMENTS

The authors wish to thank our secretary, Penny Petta, for her excellent assistance in the preparation of this manuscript.

REFERENCES

1. Alarie, Y. C. and Anderson, R. C., "Toxicologic and Acute Lethal Hazard Evaluation of Thermal Decomposition Products of Synthetic and Natural Polymers," *Toxicol. Appl. Pharm.*, *51*(2) pp. 341–362 (1979).
2. Alarie, Y. C. and Anderson, R. C., "Toxicologic Classifications of Thermal Decomposition Products of Synthetic and Natural Polymers," *Toxicol. Appl. Pharm.*, *57*(2) pp. 181–188 (1981).
3. Alarie, Y. C., Stock, M. F., Matijak-Schaper, M. and Birky, M. M., "Toxicity of Smoke During Chair Smoldering Tests and Small Scale Tests Using the Same Materials," *Fund. Appl. Toxicol.*, *3*, pp. 619–626 (1983).

4. Alexeeff, G. V. and Packham, S. C., "Use of a Radiant Furnace Fire Model to Evaluate Acute Toxicity of Smoke," *J. Fire Sciences*, Vol. 2, pp. 306-320 (1984).

5. Anderson, R. C. and Alarie, Y. C., "Screening Procedures to Recognize 'Supertoxic' Decomposition Products from Polymeric Materials Under Stress," *J. Combus. Toxicol.* 5, pp. 54-63 (1978).

6. Anderson, R. C., Croce, P. A. and Sakura, J. C., "Study to Assess the Feasibility of Incorporating Combustion Toxicity Requirements into Building Material and Furnishing Codes of New York State," Final report, Vol. I, to Department of State, Office of Fire Prevention and Control, Albany, N.Y., Arthur D. Little Inc., Ref. 88712 (1983a).

7. Anderson, R. C., Croce, P. A., Feeley, F. G. and Sakura, J. D., "Study to Assess the Feasibility of Incorporating Combustion Toxicity Requirements into Building Materials and Furnishing Codes of New York State," Final Report, Vol. II, to Department of State, Office of Fire Prevention and Control, Albany, N.Y., Arthur D. Little Inc., Ref. 88712 (1983b).

8. Anderson, R. C., Daring, K. N. and Long, M. "Study to Assess the Feasibility of Incorporating Combustion Toxicity Requirements into Building Material and Furnishing Codes of New York State," Final Report, Vol. III, to Department of State, Office of Fire Prevention and Control, Albany, N.Y., Arthur D. Little Inc., Ref. 88712 (1983c).

9. Bakhishev, G. N. [Relative Toxicity of Aliphatic halohydrocarbons to rats]. (In Russian). *Farmakol. Toxicol* (Kiev), pp. 140-142 (1973).

10. Ballantyne, B. "Relative Toxicity of Carbon Monoxide and Hydrogen Cyanide in Combined Atmospheres," *Toxicologist* 4(1), pp. 69 (1984).

11. Barrow, C. S., Lucia, H., Stock, M. F. and Alarie, Y. C., "Development of Methodologies to Assess the Relative Hazards from Thermal Degradation products of Polymeric Materials," *Am. Ind. Hyg. Assoc. J.* 40, pp. 408-423 (1979).

12. Clayton, G. D. and Clayton, F. E. (Eds), *Patty's Industrial Hygiene and Toxicology*, 34d Edition, Volume II B, p. 2941 (1981).

13. Committee on Fire Toxicology, "Fire Toxicology: Methods for Evaluation of Toxicity of Pyrolysis and Combustion Products, Report No. 2," National Academy of Sciences, Washington D.C. (1977).

14. Crane, C. R., Sanders, D. C., Endecott, B. R., Abbott, J. K. and Smith, P. W., "Inhalation Toxicology: I. Design of a Small Animal Test System. II. Determination of the Relative Toxic Hazards of 75 Aircraft Cabin Material," Report No. FAA-AM-77-9, Department of Transportation, Federal Aviation Administration, Office of Aviation Medicine, Washington D.C. (1977).

15. Gad, S. C. and Smith, A. C., "Influence of Heating Rates on the Toxicity of Evolved Combustion Products: Results and a System for Research," *J. Fire Sci. 1*, pp. 465-479 (1983).

16. Gilles, P. J. and Richard, R. W., "Toxicokinetics of [14$_C$] Hexafluoroacetone in the Rat," *Toxicol. Appl. Pharmacol.* 73, pp. 23-29 (1984).

17. Hilado, C. J. and Cummings, H. J., "A Reivew of Available LC$_{50}$ Data," *J. Comb. Toxicol* 4, pp. 415-424 (1977).

18. Hilado, C. J. and Huttlinger, P.A., "Concentration-Time Relationships for Toxic Responses," *J. Comb. Toxicol* 7, pp. 177-180 (1980).

19. Kaplan, H. L., Grand, A. F. and Hartzell, G. G., *Combustion Toxicology: Principles and Test Methods*, Technomic Publishing Co., Inc. Lancaster, PA (1983).
20. Kimmerle, M. B., "Aspects and Methodology for the Evaluation of Toxicological Parameters During Fire Exposure," *JFF/Comb. Toxicol 1*, pp. 4–51 (1974).
21. Lieu, P. J., Magill, J. H. and Alarie, Y., "An Evaluation of Some Polyphosphazenes and Commercial Fire Retardants for Wood," *J. Comb. Toxicol 9*, pp. 65–75 (1982).
22. MacFarland, H. N. and Leong, K. H., "Hazards from the Thermodecomposition of Plastics," *Arch. Env. Health. 4*, pp. 39–45 (1968).
23. MacFarland, H. N., "Respiratory Toxicology," *Essays in Toxicology*, Volume 7, Ed., W. Hayes, Jr., Academic Press, Chapter 5, pp. 121–154 (1976).
24. National Institute for Occupational Safety and Health (NIOSH). *Registry of Toxic Substances* (on line). National Library of Medicine. U.S. Department of Health and Human Services (1984).
25. Packham, S. C. and Hartzell, G. E., "Fundamentals of Combustion Toxicology in Fire Hazard Assessment," American Society of Testing and Materials, *J. Testing and Evaluation 9*(6) pp. 341–347 (1981).
26. Petajan, J. H., Voorhes, K. J., Packham, S. C., Baldwin, R. C., Einhorn, I. N., Grummet, M. L., Dinger, B. G. and Birky, M. M., "Extreme Toxicity from Combustion Products of a Fire Retardant Polyurethane Foam," *Science 187*, pp. 742–744 (1975).
27. Skog, E., "A Toxicological Investigation of Lower Aliphatic Aldehydes— I. Toxicity of Formaldehyde, Acetaldehyde, Propionaldehyde, and Butyraldehyde—As Well as of Acrolein and Crotonaldehyde," *Acta Pharmacol Toxicol 6*, pp. 299–318 (1950).
28. Wall, E. M., Presentation at the National Conference on Fire Toxicity, June 26, 1984. Arlington, VA. Sponsor, U.S. Consumer Product Safety Commission.
29. Zapp, J. A., "The Toxicology of Fire," Medical Division Special Report No. 4, U.S. Army Chemical Center, Maryland p. 92 (1951).

Combustion Conditions and Exposure Conditions for Combustion Product Toxicity Testing

CLAYTON HUGGETT

Center for Fire Research
National Bureau of Standards
Gaithersburg, MD 20899

ABSTRACT

A number of procedures have been described in the literature for investigation of the inhalation toxicity of combustion products. There is need for agreement on test methods and test conditions to facilitate communication, allow the exchange of data, and provide a basis for control of hazards due to combustion products in fires. Combustion systems and animal exposure systems which have been employed are classified according to their basic attributes. Simple considerations of limiting stoichiometry in the combustion module can guide the selection of conditions which simulate real fire environments. The dynamics of the exposure system will determine the procedural dose received by the test animal and can be related to real fire exposure. Many past investigations have failed to take adequate account of these fundamental principles.

INTRODUCTION

A NUMBER OF experimental procedures have been described in recent years for the investigation of the inhalation toxicity of combustion products [1,2]. These vary widely in operational detail, making comparison of results between laboratories difficult. There is need for agreement on test methods and test conditions to facilitate communication between researchers, to allow the exchange of data, and to provide information useful for control of the hazard due to combustion products. This paper addresses some of the conditions affecting combustion of the specimen and exposure of experimental animals that have been frequently ignored but that must be considered in selecting acceptable test methods.

The procedures reported in the literature have been variously described as being intended for research, to facilitate communication, for screening, to rate or rank materials, for comparison, for listing, to provide data for modeling, for use in a hazard assessment, for use in standards and specifications, for regulation, and for other purposes. Obviously, these varied purposes may require different types of tests. The first step in selecting or developing a test method must be to define the purpose for which the test is to be used. Once the purpose has been precisely described, care must be exercised to see that the procedure is not unduly elaborated or compromised to accommodate other peripheral purposes.

The primary purpose of a combustion product toxicity test should be to provide quantitative information on the toxicity of combustion products that can be used along with other fire properties and fire system properties to judge the suitability of a material or product, from a fire safety standpoint, for a specific application.

Numerous authors have discussed the attributes that are desirable in a standard fire test method. These range from philosophical discussion to fairly detailed requirements. A number of them deal specifically with combustion product toxicity testing. Birky [3] has discussed the philosophy of testing of combustion products. Packham and Hartzell [4] discussed the toxicological aspects of the test procedures. A committee of the National Academy of Sciences [5] studied the problem of combustion product toxicity testing and made detailed recommendations concerning test methodology. ISO/TC92/SC3 has had extensive discussion of toxicity test requirements. Their report no. 6543 [6] is now being revised and expanded. However, it deals largely in generalities and provides little specific guidance for choosing among different procedures. The British Standards Institute has also published recommendations on the development of a toxicity test method [7].

Recently Anderson, Croce, and Sakura [8] evaluated many of the published test procedures against a set of criteria which they selected. Their purpose was to recommend "the most useful test for evaluation of toxic potency of smoke" for use by the state of New York. The rationale for their criteria is discussed in their report.

All of the published test methods involve a method of thermally decomposing* or combusting the material under investigation, followed by characterization of the properties of the volatile combustion products by analytical procedures and/or observation of their physio-

---

*For convenience I will frequently refer to this process as combustion and to the products as combustion products even though in some cases the thermal decomposition process may not involve oxidative combustion.

logical effects on laboratory test animals. Numerous variations in apparatus and operating procedures have been employed in order to simulate various characteristics of smoke generation** and the exposure of victims in real fires. Many of these variations are summarized in Table 1. Selection of one or more features from each column in the table will define a test method. All the test methods described in the literature can be classified in this manner and it is found that no two are exactly alike. It is apparent that hundreds of combinations of features are possible, each of which would define an operable test method. A task of the fire scientist is to select from this large number of possibilities the test method or methods which best simulate the conditions encountered in real fires.

Fire statistics are of limited help in selecting conditions for toxicity testing. There is evidence that a large number of fatal fires start in the smoldering mode, but these usually undergo a transition to a flaming mode at a later stage and it is not clear to what extent incapacitation or death occurs during the smoldering phase. British data indicate that a majority of casualties occur in the room of fire origin [9] while American data suggest that a large number of casualties occur outside the room of origin [10]. Whether these differences reflect differences in types of fires, building design, or methods of recording data is not clear at this time. It appears, however, that smoldering combustion, free burning fires, and ventilation limited fires are all of importance in producing casualties from smoke inhalation and should be simulated in toxicity test methods.

Table 1. Characteristics of test methods.

Combustion Mode	Sample Exposure	Animal Exposure	Assessment
Pyrolysis	Cup Furnace	Static	Chemical Analysis
Flaming	Radiant Furnace	Steady State	Biological
Free Burning	Free Flame	Time Dependent	Incapacitation
Ventilation Limited	Tube Furnace	Dynamic	Death
Smoldering	Static	Steady State	Respiratory Rate
Glowing	Flow	Time Dependent	Blood Chemistry
	Traveling	Whole Body	Delayed Effects
	Co-Current	Nose Only	EC_{50}
	Counter Current	Instrumented	Time to Effect
	Autoignition		Critical Temperature
	Piloted Ignition		Critical Concentration

**The ASTM E176 definition of smoke as "the airborne solid and liquid particulates and gases evolved when a material undergoes pyrolysis or combustion" is used here since all of these components may contribute to the toxic effect observed.

While a test method based on chemical analysis of the combustion products might ultimately be desirable, there is general agreement that at the present time some form of bio-assay is necessary to deal with the complexity of the product mixture and the resulting physiological effects. The assessment of physiological effects of combustion products on experimental animals has been discussed recently by Kaplan and Hartzell [11] and will not be discussed further here.

COMBUSTION PRODUCT GENERATION

The composition, quantity, and rate of formation of combustion products will depend on a number of factors in addition to the chemical composition of the sample. Chief among these will be the physical state of the sample, the ventilation conditions, and the energy flux to the sample. Our principal concern will be with solid fuels, usually natural or synthetic polymers, which make up the bulk of the fuels contributing to unwanted fires.

Attempts have been made to demonstrate the relationship of the composition of the combustion products from various laboratory systems to those produced in large fires. These studies have included measurements of the concentration of major combustion products in the laboratory and in large scale fire experiments, comparison of CO/CO_2 ratios, comparison of profiles of minor products by GC/MS techniques (fingerprinting), and exposure of test animals in large scale fire experiments. While much useful information has been obtained, these techniques suffer from shortcomings which limit their use as a guide to the selection of laboratory combustion procedures. It is not possible to simulate on a large scale more than a few of the large number of possible fire scenarios. Combustion product concentrations in large fires are highly nonuniform and the location of the areas of toxic threat with respect to the source of the fire gases is uncertain. The large scale experiments are time-consuming and costly. In view of these difficulties it may be useful to consider an alternative approach to the selection of laboratory combustion and exposure conditions based on a consideration of the fundamental characteristics of the combustion process.

Modes of Combustion

Four general modes of thermal decomposition can occur in fires; pyrolysis, flaming combustion, glowing combustion, and smoldering combustion. The nature of the gaseous combustion products formed can differ drastically between modes. Within a given mode, environmental conditions (ventilation and heat flux) can further moderate the composition of the products.

Pyrolysis

When an organic solid is heated by an external heat source it will begin to decompose when its temperature reaches approximately 400 °C. This relatively constant decomposition temperature is a consequence of the fact that the decomposition usually involves the breaking of carbon-carbon or carbon-oxygen bonds and these have similar bond strengths regardless of the detailed chemical environment. In the case of linear polymers, the initial decomposition may take the form of chain breaking or unzipping, resulting in the formation of volatile fuel fragments. A classic example is poly (methylmethacrylate) (PMMA) which, at a temperature of about 400 °C, is converted almost quantitatively to monomer. Because of the high activation energy of the decomposition process, an increase in energy flux to the sample will result in an increase in the rate of decomposition but will have very little effect on the surface temperature or the nature of the decomposition products. Few polymers unzip as cleanly as PMMA but many will pyrolyze completely at a relatively constant temperature to give a mixture of volatile combustible products.

Cross linked polymers, many natural products, and polymers containing polar side groups on the chain may decompose by the elimination of small molecules, leading to further cross linking and the formation of a carbonaceous char layer on the surface. The volatile products may be relatively non-combustible due to the presence of large proportions of molecules such as H_2O, HCl, and CO_2. The char layer will insulate the underlying polymer but the surface temperature will increase because of the greater thermal stability of the char. The pyrolysis wave will propagate into the sample, maintaining a fairly constant temperature at the pyrolysis front, and the product vapors will escape through the porous char layer. If the surface temperature of the char reaches a high enough value, surface oxidation of the char may occur (glowing combustion) leading to further energy release and, perhaps, the initiation of flaming combustion. The volatile pyrolysis products can mix with air to form a flammable mixture leading to flaming combustion under favorable circumstances. Transition from pyrolysis or smoldering to flaming cannot be predicted with certainty.

The autoignition temperature of the fuel-air mixture will usually be above the pyrolysis temperature so flaming combustion may not occur in the absence of a high temperature ignition source (flame, spark, or glowing char). Most fires which provide sufficient energy flux to pyrolyze solid fuels will also provide pilot ignition sources so pyrolysis gases produced in the absence of flaming combustion will be of little toxicological consequence in most cases. An exception might be the case where the fuel is shielded from direct contact with the flame, for example, when used as insulation in a wall cavity. Ureaformaldehyde foam is sometimes used in such application and is known to give off large quantities of HCN when strongly heated. A fire confined within an insulated room might produce enough heat to pyrolyze the insulation and generate sufficient toxic pyrolysis products to pose a threat elsewhere in the building.

Flaming Combustion

If a pilot ignition source is present or the autoignition temperature is exceeded, flaming combustion will start and continue as long as the fuel-air mixture remains within the flammable limits. In this case the composition of the combustion products will be strongly influenced by the oxygen supply.

In a well-ventilated fire reaction will be nearly complete, heat release will be at a maximum, and the principal products will be relatively nontoxic CO_2 and H_2O. If the oxygen supply is limited, combustion will be incomplete leading to the survival of CO and a variety of intermediate organic products. The toxic potency of this smoke may be much greater than that of the completely oxidized combustion products. Obviously, the oxygen supply can vary widely during the course of a fire depending on the fire intensity and ventilation condition [12]. Conditions encountered in a fire may cover a continuum ranging from a fully-ventilated fire during the ignition phase to a ventilation limited fire where the oxygen concentration drops to a very low value, preventing flaming combustion.

Test conditions cannot hope to simulate all of these conditions so it will be necessary to select a limited number of conditions which simulate the most important aspects of the fire phenomena. Fires involving flaming combustion are the most prevalent among hostile fires but they are also the most variable with respect to the composition of products of combustion. It appears that at least two conditions must be employed in a method to assess the toxicity of combustion products from flaming combustion; a well-ventilated combustion regime and one in which the oxygen supply is severely limited leading to the formation of products of incomplete combustion.

Smoldering Combustion

Certain porous fuels such as cotton batting, sawdust, or some plastic foams can undergo smoldering combustion [13]. The necessary conditions are a porous fuel bed, allowing access of air, and good thermal insulation in the fuel to conserve the relatively low heat output of the smoldering reaction. Smoldering takes place as a self-propagating wave in a porous fuel bed leaving behind a char residue. Temperatures are relatively low in the smoldering front and the products of smoldering are under-oxidized organic fragments and a carbonaceous char residue. Because of the low temperature and small fuel output, smoldering can continue for long periods of time without breaking forth into flaming combustion.

Pure smoldering combustion is relatively uncommon compared to flaming combustion but fires which originate in the smoldering mode show an unusually high fatality rate. Smoldering occurs in upholstered furniture or bedding, frequently initiated by carelessly discarded smoking material, and can generate large volumes of toxic smoke while the occupants are asleep. Other common incidents involving smoldering combustion are the smoldering of insulation initiated by faculty electric circuits or the smoldering of particle board initiated by careless workmen. Smoldering is an insidious hazard because it provides little warning of its presence in the absence of a smoke detector and because it can persist for hours at a very low level, producing toxic products and breaking out into a flaming fire long after the initial ignition source has disappeared. It should be recognized that none of the existing toxicity test methods satisfactorily simulate the conditions of smoldering combustion.

Glowing Combustion

The combustion of a cellulosic fuel or a cross linked polymer may result in the formation of a carbonaceous char layer on the burning surface. This may slow the evolution of flammable fuel and flaming combustion may cease. Oxygen will diffuse to the char surface where a surface oxidation may take place, producing large amounts of CO. The glowing combustion of charcoal in a picnic grill is a classic example of this mode of combustion. Such devices have sometimes been used as heating units in confined spaces with tragic results due to the generation of lethal concentrations of CO. Such conditions may occur late in a fire, after active flaming has stopped and charred debris remains. It may be present during overhaul operations where unsuspecting firemen may remove their breathing apparatus and be exposed to odorless CO. Glowing combustion is probably of lesser importance to fire safety because conditions conducive to glowing combustion are seldom encountered during the active phase of accidental fires.

EXPERIMENTAL COMBUSTION CONDITIONS

Flammability Limits

It is well known that the composition of a gaseous fuel-oxidizer mixture must lie within certain limits for a combustion wave to propagate. It is difficult to apply this concept to the diffusive burning of a solid fuel. Nevertheless, it is possible to estimate a lean "flammability limit" below which flaming combustion cannot take place.

When a solid fuel undergoes flaming combustion, it first decomposes to form volatile fuel which mixes with air to form a flammable mixture. The exact composition of the mixture may be unknown but it must have a combustion energy content in excess of 1.8 kJ/l at standard conditions in order to support a flame [14]. With a hydrocarbon fuel such as polypropylene, this corresponds to a minimum loading to produce a flammable mixture of approximately 40 mg/l. This is of the same order of magnitude as the LC_{50} in typical toxicity tests. For fuels containing oxygen or hetero-elements or fuels which do not volatilize completely the critical loading will be much higher while the LC_{50} may be much lower.

Since for a material producing highly toxic combustion products this critical loading may exceed the lethal concentration by one or more orders of magnitude, care must be taken to see that the fuel concentration exceeds the lower flammability limit if flaming combustion is to be simulated. This has not been the case in some of the procedures described in the literature. A mixture whose composition lies within the flammability limits may be achieved by confining the combustion reaction to a relatively small volume and allowing the combustion products to mix with additional air in a large chamber, as in the NBS procedure [15], or by diluting the product stream with secondary air in a flowing stream, as in the DIN procedure [16].

Ignition

When a sample of a combustible fuel is exposed to a heat source, as in a ramp heating experiment or when introduced into a preheated furnace, volatile fuel gases will be generated and will mix with the air. These gases will be carried away from the fuel source by diffusion and convection and, if the sample is small, may never reach a concentration within the flammable limits where active flaming can take place. Thus, the so-called autoignition temperature has been found to depend on sample size and apparatus [15]. Even when the flammability limits are exceeded flaming combustion may not occur because the autoignition temperature is not reached before the fuel supply is exhausted.

Flaming combustion may be initiated in a flammable mixture if the temperature reaches a high enough value to initiate a self sustaining oxidation reaction (autoignition) or by a small high temperature source such as a spark or flamelet (piloted ignition). Piloted ignition sources are usually present in fires so pyrolysis in the absence of flaming ignition will seldom be encountered. Smoldering ignition will be the significant exception. However, it is not clear that many of the combustion modules used in toxicity experiments can assure flaming combustion under conditions of variable sample size, heating rate, and ventilation. A positive ignition source should always be provided in cases where flaming combustion is to be simulated.

Stoichiometry

Obviously, the ratio of fuel to oxygen in the combustion module will have a great effect on the mode of combustion and the composition of the products. Conditions in the combustion module are seldom precisely defined in combustion toxicity experiments and they may change during the experiment as fuel and oxygen are consumed and heat is liberated. Nevertheless, it is possible to draw conclusions concerning the limits of the combustion process from simple stoichiometric considerations.

The decrease in oxygen concentration can lead to incomplete combustion with a resulting shift in the composition of the combustion products and it can produce physiological effects on the experimental animals due to anoxia. The latter effect can sometimes be avoided by the introduction of secondary air or oxygen after combustion has ceased, but the former effect is difficult to avoid and indeed is characteristic of real fires. These effects will be continuous functions of the oxygen concentration, but it is generally conceded that when the oxygen content of the air has fallen to two-thirds its normal value (from 21% to 14% by volume) significant effects on both combustion behavior and physiological response will be observed.

One liter of air at standard conditions will contain approximately 0.27g oxygen. The oxygen demand for complete combustion to CO_2 and H_2O will range from about 3.4g oxygen per gram of a hydrocarbon fuel such as polypropylene to about 1.19g oxygen per gram of a highly oxygenated fuel such as cellulose. Such figures assume complete consumption of oxygen present. In practice at least three times the amount of oxygen would have to be furnished to sustain flaming combustion. Thus, a minimum of 4 to 12 liters of air will be required per gram of sample to provide conditions simulating a poorly ventilated fire. A much larger volume of air will be required to simulate well ventilated conditions.

Typical Test Conditions

We can now examine the combustion environment in some typical toxicity tests.

In the NBS test [15] samples of up to 7 or 8 grams are thermally decomposed in a cup furnace having a volume of approximately one liter. The furnace communicates freely with a 200l exposure chamber but the degree of mixing of air in the furnace is not known. It seems certain that much of the initial decomposition is a pyrolysis reaction which takes place under conditions of extreme oxygen deprivation. Secondary combustion of the fuel vapors may or may not occur at the mouth of the furnace where the oxygen supply is presumed to be adequate. The size of the sample used in the test to produce the desired physiological effect varies widely from material to material. The sample size is varied for a single material to generate a dose-response relationship, further confusing an attempt to define the fuel-air ratio. The ratio will change during the course of an experiment as fuel and oxygen are consumed and fresh air enters the furnace in an uncontrolled manner. It is apparent that this system fails to carry out the combustion in a well-defined manner.

The combustion module used in the DIN 53436 procedure provides what is perhaps the best controlled combustion environment of any toxicity test combustion module described in the literature [16]. A tube furnace moves along a tubular combustion vessel at a constant rate, providing a constant fuel supply to an air stream flowing counter current to the direction of furnace travel. This gives a nearly constant concentration of combustion products in the exit stream over the duration of the experiment. While standard operating conditions have not yet been specified and can be varied with the experiment, a sample feed rate of 0.12g/min. and an air flow rate of 3.33l/min. appear typical. This will give a "concentration" of 36mg/l in the effluent stream without dilution, in the range of the 30 minute LC$_{50}$ value for many fuels. Such conditions would be expected to support flaming combustion in a restricted oxygen supply. The degree of oxygen depletion will vary from sample to sample, but a lower limit to the oxygen concentration can be calculated for any experiment. The flow of air counter to the direction of flame propagation differs from that encountered in accidental fires. Particularly with porous fuel samples, this flow pattern may affect the structure of the combustion zone and the composition of the products [17]. The apparatus is capable of covering a range of stoichiometries under closely controlled conditions. The relationship of combustion in a counter-current tube furnace to that encountered in real fires has not been established.

Alarie and Anderson [18,19] developed a test method at the University of Pittsburgh in which the sample, usually 2 to 50g, is heated at a rate of 20°C/min in a 20l box furnace while an air stream of 11l/min passes through the furnace and carries the products to the animal exposure chamber. Obviously, the fuel/air ratio can vary widely depending on the weight of sample and its rate of combustion, ranging from well ventilated to oxygen limited conditions. The rate of sample weight loss is monitored so it is possible to estimate conditions in the furnace at any time. After the sample ignites, it burns with a diffusion flame, which may simulate conditions encountered in real fires better than some other combustion schemes.

In the dome chamber method [20] a one gram sample is heated at a rate of 40°C/min. in a closed end tube attached to a 4.2l animal exposure chamber. It is apparent that complete combustion of the sample would exhaust the oxygen in the chamber. In practice, the oxygen content of the chamber air is found not to fall below the physiological limit so only a small part of the sample has undergone oxidative combustion. As the sample is heated rapidly, it reaches a temperature at which it starts to decompose and may catch fire. The oxygen in the heating tube is quickly exhausted and oxygen from the exposure chamber may be consumed more slowly as it diffuses into the furnace tube. Active flaming ceases but since the heating is continued to 800 °C the bulk of the sample will be pyrolyzed under extremely oxygen deficient conditions. It is difficult to relate this combustion module to conditions in real fires.

This type of analysis can be applied to any of the combustion modules described in the literature, but the four considered here are representative of most of their significant variations. The environmental conditions in most of the combustion modules that have been used for combustion product toxicity testing are not sufficiently well defined to permit close analysis but limits can be placed on their performance. It can be shown that the combustion conditions employed in a number of these devices sometimes bear little resemblance to conditions encountered in real fires.

ANIMAL EXPOSURE CONDITIONS

Classification of Exposure Modules

Animal expsoure systems are also classified as static or dynamic. Obviously, such classifications are closely related to the operation of the combustion system. If combustion occurs in a closed system and the animals are exposed to the total products of combustion, the exposure system is classified as static. If air flows continuously through the combustion chamber and the products are transported through the animal exposure chamber in a flowing stream, the system is considered to be dynamic.

Such classification is of limited usefulness in characterizing the conditions of animal exposure because of wide variations in exposure conditions possible within these broad classifications. A more useful procedure is to consider the time-concentration profile of combustion products to which the animals are exposed.

The DIN 53436 procedure is a dynamic system which exposes the animals to a nearly constant concentration of combustion products during the entire exposure period [16]. The NBS procedure, a static system, exposes the animals to a similar nearly constant concentration during much of the exposure period [15]. An initial transient exposure while the sample is decomposing is short compared to the 30 minute exposure period. This initial transient can be largely avoided by a modification now under investigation in which sample combustion is completed before the animals are exposed [21]. Some decrease in concentration during exposure can be expected in static systems due to animal uptake, settling, and deposition on chamber walls. We would expect the dynamic DIN procedure and the static NBS procedure to produce similar effects when the animals are exposed to similar concentrations of combustion products.

On the other hand the University of San Francisco procedure (Dome Chamber Method), a static method in its most widely used configuration, exposes the animals to a steeply rising concentration gradient until death occurs [20]. The University of Pittsburgh method, a dynamic method, also exposes the animals to a varying concentration of combustion products, first increasing and then decreasing as combustion is completed and the products are swept from the furnace and exposure chamber [18,19]. It is of interest to consider how the animals would be expected to respond to these different exposure profiles.

Dose-Response

Classic inhalation toxicology is based on the establishment of a dose-response relationship where physiological response is related to the amount of toxicant (dose) received by the animals. Procedural dose is defined as the product of concentration and time of exposure. This differs from the biological dose, the quantity of toxicant actually inhaled, which is usually unknown. In pure gas studies, where concentration can be precisely measured and maintained, application of the procedural dose concept is relatively straightforward. In combustion toxicology, where both concentration and composition can vary with time and the measurement of concentration is difficult, application of the dose-response relationship is more difficult. Nevertheless, it appears to offer the best method now available for correlating lethality producing physiological effects with smoke exposure [4].

The physiological effect of the combustion products is often described by giving the EC_{50} the concentration of products necessary to produce the observed effect in 50 percent of the test animals. Such a description is obviously incomplete since it does not include the time of exposure. An exposure time of 30 minutes has been widely used and is often implied when EC_{50} values are quoted without reference to exposure time. Toxicants which produce narcosis or death usually show a response which depends on cumulative dose, and the time of exposure must be explicitly included. The results are properly reported as the effective dose, ED_{50}. Irritants which affect respiratory behavior, on the other hand, produce a response which depends on concentration and the use of an EC_{50} may be appropriate in this case [22].

For procedures which expose the animals to relatively constant concentrations of combustion products for a fixed period of time, as in the DIN procedure or the modified NBS procedure, the simple time-concentration product, Ct, provides a useful measure of procedural dose within the concentration range of interest (from the threshold response concentration to concentrations causing rapid death). It should not be assumed, however, that the dose to produce a specified effect is independent of concentration as required by Haber's law even within these limits. The observed effect will be related to the amount of toxicant absorbed by the animal, not to the atmospheric concentration (the biological dose as distinguished from the procedural dose). Exposure to irritants may decrease the respiratory volume, thus slowing the inhalation of toxic products. Toxicants which are absorbed reversibly, such as CO, will approach an equilibrium concentration in the body which is dependent on the atmospheric concentration. Even in the case of toxicants which are absorbed irreversibly, such as HCN, metabolic processes may remove some of the toxicants, thus distorting the dose-response relationship.

When the animals are exposed to a smoke concentration which varies with time, as when the products are generated by a ramped temperature increase in the combustion module, the problem is even more complex. As a first approximation, the dose may be taken to be equal to the integral of the concentration over the exposure time. (It is assumed that the animal's respiration rate is rapid compared to the rate of change in concentration in the atmosphere so the inhaled gas is representative of the time-concentration profile.)

Where the concentration of combustion products varies with time, the procedural dose may be defined as:

$$dose = \int_{t_{\infty}} C_{(t)} dt \qquad (1)$$

For the case where the sample is decomposed in a rapidly moving air stream with little internal mixing (plug flow) we have:

$$dose = \int_{t_{\infty}} \frac{\dot{m}}{\dot{v}} dt \qquad (2)$$

where $\dot{m}$ is the rate of mass loss of the sample and $\dot{v}$ is the volumetric flow rate (usually constant). When the products of decomposition are swept from the exposure chamber during the experiment, the dose becomes simply:

$$dose = \frac{\Delta m}{\Delta v} t_{\infty} = \frac{\Delta m}{\dot{v}} = \frac{\Delta m}{V} \tau \qquad (3)$$

where Δm is the total mass of sample converted to smoke, Δv is the total flow volume during the exposure, V is the volume of the exposure chamber, and τ is the average residence time of the gas in the exposure chamber ($\tau = V/\dot{v}$). Fresh air may be mixed with the effluent from the combustion chamber before entering the exposure chamber to vary the concentration of products, reduce the temperature and increase the oxygen content. The residence time will be decreased by the diluting air and the procedural dose to which the animals are exposed will be decreased accordingly.

For a more complex flow pattern equation (2) may be integrated by numerical means if $\dot{m}$ and $\dot{v}$ are known. A further complication arises in dynamic systems if either the furnace or exposure chamber has a significant volume so the assumption of plug flow is no longer reasonable ($\tau \rightarrow t_{ae}$). In this case, internal mixing of the combustion products and the air stream will occur to an indeterminate extent. A limiting case will occur if the sample is assumed to decompose instantaneously in a furnace chamber of volume V_f to give an initial concentration $C_o = \Delta m/V_f$ and then be swept out by the air stream with perfect mixing. Because of the steep temperature rise and high activation energy characteristic of most systems, the assumption of instantaneous decomposition (short compared to the exposure time) will be a good approximation. In this case the concentration of combustion products in the exit stream from the furnace becomes the input concentration to the exposure chamber (with or without further dilution with fresh air) and the procedural dose is estimated accordingly. Under these conditions, the concentration in the furnace chamber and in the exit stream will decrease exponentially and the procedural dose to the exposure chamber will be given by:

$$dose = C_o \int_{t_{ae}} e^{t/\tau}\, dt = C_o\tau(1 - e^{t/\tau}) \tag{4}$$

When τ is small compared to the exposure time, the dose approaches the limiting value of $C_o\tau$. When τ is large the dose approaches the static value $C_o t$.

We would expect that a static or a dynamic exposure mode would give approximately the same dose response, within the range of concentrations where Haber's law is a reasonable approximation, if proper allowance is made for the concentration profile during exposure. Unfortunately, this has not been done in the past and no valid comparison of the two systems has been made.

Comparison of Exposure Methods

Anderson et al [8] have recently made a comparison of the results of toxicity measurements made by a dynamic method similar to the University of Pittsburgh method and the static NBS method. They take as their measure of toxic potency the total mass of sample charged to the test system to produce the desired physiological effect. They find good numerical agreement between the two methods for a variety of thermoplastic materials but for a number of char forming materials, primarily cellulosics, the University of Pittsburgh test indicated a lower level of toxicity by approximately a factor of three.

An examination of the experimental procedure indicates that in their method the samples were decomposed in a 42l box furnace with an air flow of 11l/min. While no mixing is provided, we can assume that diffusive and convective mixing will result in a fairly uniform product distribution in the furnace. The sample is heated at a rate of 20 °C/min and the thirty minute animal exposure period starts with the start of decomposition. It is reasonable to assume that decomposition occurs early in the exposure period. The system will approximate the well stirred case described by Equation 4. Since the residence time is short compared to the exposure time (4 min vs. 30 min) the products will be almost completely swept out of the oven during the exposure. The combustion products are diluted with an additional 9l/min of fresh air before passing through a 2.2l exposure chamber. The procedural dose to which the animals are exposed is then approximately 0.05g min/l per gram of sample. In the NBS method, the sample is decomposed in a static chamber of 200l volume. Again, the exposure period is 30 min so the procedural dose is approximately 0.15g min/l per gram or three times as great as in the Pittsburgh test for equal weight of sample. Thus, when the comparison is made on the basis of procedural dose, the two procedures agree reasonably well for char forming materials but the Pittsburgh results for thermoplastics are higher than the NBS results by approximately a factor of three.

It is interesting to speculate on possible differences between the results of the two procedures. In the static procedure, the sample is exposed to a fixed high temperature and is decomposed in a time which is short compared to the animal exposure period. The animals are exposed to a uniform mixture of all of the decomposition products for most of the exposure. In the dynamic procedure, on the other hand, the sample is exposed to a ramped temperature increasing at a rapid rate. Decomposition occurs over a period of several minutes at increasing temperatures and the composition of combustion products to which the animals are exposed may vary with time and temperature in the dynamic exposure chamber. Cellulose products, for example, begin to decompose at a relatively low temperature, giving off water vapor and volatile organic compounds which are likely to be irritants of relatively low toxicity. A carbonaceous char is formed which, at higher temperatures, may undergo glowing combustion to give large amounts of CO. Polyurethanes first break down to give isocyanates which are highly irritating and then produce hydrogen cyanide at higher temperatures. Many thermoplastics, on the other hand, volatilize at a relatively fixed temperature to give a homogeneous mixture of combustion products. Other thermoplastics such as polyvinylchloride may eliminate small polar molecules at low temperatures, followed by organic products and CO at higher temperatures. It may be noted that all the thermoplastics investigated by Anderson et al contain Cl, F, or N. It would be of interest to make comparisons with thermoplastics such as polyethylene or polystyrene where the elimination of polar molecules does not occur.

If the animals are first exposed to irritant gases in a flow system, their respiration rates and tidal volumes will be reduced. This may affect the rate of uptake of other toxicants at a later stage of exposure. Early exposure to a narcotic gas can be expected to have an opposite effect. Consequently, differences in dose response relationships between steady state exposure and ramped heating exposure are not unexpected. A careful study of such differences, accompanied by time resolved analytical and respiratory data, may provide a useful method of studying the physiological effects of complex combustion product mixtures.

It may also be useful to consider which situation is more representative of real fire scenarios. It may be noted that in the case of materials of layered construction, such as wall lining or floor covering materials, which would be exposed to a heat source from a fire from only one side in normal application, the composition of combustion products can be expected to vary with time. Such an exposure is simulated by the Weyerhaueser radiant furnace modification of the NBS chamber [23]. A flowing exposure system might be useful in this situation to separate the effects of products from successive layers of the sample.

Other investigators have attempted similar comparisons of static and dynamic tests. Cornish et al measured the toxicity of combustion products from the same materials in a static and a dynamic system [24]. They found the static system to require a roughly ten-fold greater weight of sample to cause death of the animals than the dynamic system. A review of their experimental procedures indicates that in the dynamic procedure the procedural dose per gram of material was approximately twice that of the static method, but a factor of five between the two methods remains unexplained. They note that CO producing materials are relatively more toxic in the static chamber and that halogen halide producing materials are more toxic in the dynamic chamber. It was suggested that this difference was due to a decrease in the concentration of polar compounds during long exposure in the static chamber. Another possible explanation is that the relatively short residence time during the period of active decomposition of the sample in the dynamic system produces a high concentration of combustion products for a relatively short time, resulting in conditions where Haber's rule does not apply.

Lawrence, Page, Singh, and Autian compared results of a dynamic method developed at the University of Tennessee with a static method from NASA/JSC [25]. They state, "It was somewhat unexpected to find a greater percentage of exposed rats were killed by pyrolysates from the (dynamic) procedure, per unit weight of sample, than from the (static) procedure." A stirred 63*l* exposure chamber and a flow of 1.5*l*/min of air was used in the dynamic system. With the long residence time, most of the combustion products will be retained in the exposure chamber during the 60 minute exposure period so the dose can be approximated by Equation (4). It is found that the procedural dose per unit weight of sample received by the animals in the dynamic chamber is about 2 1/2 times the dose received in the static test. Making this adjustment, the two methods are in much better agreement although some discrepancies remain.

Hilado attempted a comparison of static and dynamic systems using modifications of the Dome Chamber method [26]. He measured time to effect using massive samples and ramp heating. The concentrations of combustion products are not defined so no valid comparisons of toxic potency are possible. His results are best interpreted in terms of the thermal stability of the fuel.

There appears to be no a priori reason for the selection of static or dynamic exposure conditions. Either can be expected to give a valid measure of the physiological response of the test animals under the test conditions selected if the procedural dose is properly evaluated. Some differences in response can be expected due to differences in the time sequence of the generation of combustion products. In real fires, this variation in the composition and concentration of combustion products may be due to the successive ignition of different fuel elements, the burning of a layered product such as a wall covering or carpet, or changes in temperature and ventilation as the fire progresses. The exposure conditions leading to casualties may range from dynamic conditions in close proximity to a developing fire to a more integral exposure in a compartment remote from the scene of fire origin. Consideration of the expected behavior of the product under use conditions and the dynamics of the test system can guide the selection of meaningful test conditions.

CONCLUSIONS

It is desirable to select a limited number of test conditions from among the multiplicity that have been or could be proposed for combustion product toxicity testing in order to promote communication, allow the comparison of data from different sources, and provide a basis for control of the hazard due to combustion products.

Conditions selected for the testing of a material or product must reflect the expected use environment and the likely fire exposure environment.

Simple considerations of limiting stoichiometry in the combustion module can guide the selection of conditions which simulate expected conditions in real fires.

The dynamics of the exposure system will define the procedural dose received by the test animals and can be used to relate laboratory test results to expected human exposure.

REFERENCES

1. Clarke, F. B., "Toxicity of Combustion Products: Current Knowledge," *Fire Journal*, p. 84 (September 1983).
2. Kaplan, H. L., Grand, A. F., and Hartzell, G. E., "Combustion Toxicology: Principles and Test Methods," Technomic Publishing Company, Inc., Lancaster, Pennsylvania (1983).
3. Birky, M. M., "Philosophy of Testing for Assessment of Toxicological Aspect of Fire Exposure," *J. Combustion Toxicology*, 3, pp. 5-23 (1976).
4. Packham, S. C., Hartzell, G. E., "Fundamentals of Combustion Toxicology in Fire Hazard Assessment," *J. Testing and Evaluation*, 9, pp. 341-347 (1981).
5. Fire Toxicology: Methods for Evaluation of Toxicity of Pyrolysis and Combustion Products, Report. No. 2, Committee on Fire Toxicology, National Research Council, National Academy of Sciences, Washington, DC (August 1977).
6. The Development of Tests for Measuring Toxic Hazards in Fire, ISO/TR 6543-1979(E), ISO/TC92.
7. Report and Recommendations on the Development of Tests for Measuring the Toxicity of Combustion Products, PD 6503:1982, British Standards Institute, London.
8. Anderson, R. C., Croce, P. A., Sakura, J. D., "Study to Assess the Feasibility of Incorporating Combustion Toxicity Requirements into Building Materials and Furnishings Codes of New York State," Arthur D. Little, Inc., Final Report to Department of State, Office of Fire Prevention and Control, Albany, New York (May 1983).
9. Harland, W. A. and Anderson, R. A., "Causes of Deaths in Fires;" in "Smoke and Toxic Gases from Burning Plastics," QMC Industrial Research, Ltd., 229 Mile End Road, London E14AA (January 1982).
10. Gomberg, A., National Bureau of Standards, Private Communication Based on Analysis of NFIRS data for 1979-1982.
11. Kaplan, H. L. and Hartzell, G. E., "Modeling the Toxicological Effects of Fire Gases: I. Incapacitating Effects of Narcotic Fire Gases," *J. Fire Sciences*, 2, pp. 286-305 (1984).
12. Tewarson, A. and Steciak, J., "Fire Ventilation," *Combust. Flame*, 53, pp. 123-134 (1983).
13. Ohlemiller, T. J. and Lucca, D. A., "An Experimental Comparison of Forward and Reverse Smolder Propagation in Permeable Fuel Beds," *Combust. Flame*, 54, pp. 131-147 (1983).
14. Huggett, C., "Flash Fire Hazards in Fire Experiments," *J. Fire Sciences*, 1, pp. 396-398 (1983).

15. Levin, B. C., Fowell, A. J., Birky, M. M., Paabo, M., Stolte, A. and Malek, D., "Further Development of a Test Method for the Assessment of the Acute Inhalation Toxicity of Combustion Products," NBSIR 82-2532, National Bureau of Standards (June 1982).
16. Klimisch, H. J., "The DIN 53436 Toxicity Test Method, in Smoke and Toxic Gases from Burning Plastics," QMC Industrial Research Ltd., 229 Mile End Road, London E14AA (January 1982).
17. Herpol, C. and Vandevelde, P., "Effect of Furnace Direction on Toxicity Results in a Dynamic Moving Tubular Furnace Method," *J. Combust. Toxicology*, 9, pp. 165-186 (1982).
18. Alarie, Y. C. and Anderson, R. C., "Toxicologic and Acute Lethal Hazard Evaluation of Thermal Decomposition Products of Synthetic and Natural Polymers," *Toxicol. Appl. Pharm.*, 51(2):341-362 (1979).
19. Alarie, Y. C. and Anderson, R. C., "Toxicologic Classification of Thermal Decomposition Products of Synthetic and Natural Polymers," *Toxicol. Appl. Pharm.*, 57(2):181-188 (1981).
20. Hilado, C. J., "Proposed Standard Test Method for Relative Toxicity of Smoke Evolved from Pyrolysis or Combustion of Materials by the Dome Chamber Method," ASTM E5.21 (March 1983).
21. Levin, B. C., National Bureau of Standards, Private Communication.
22. Alarie, Y., "Sensory Irritation by Airborne Chemicals," *CRC Crit. Rev. Toxical.*, 2, p. 299 (1973).
23. Packham, S. C., "Proposed Standard Method for Testing Performance of Materials in a Radiant Furnace Using a Biological Assay," ASTM E5.21 (October 1983).
24. Cornish, H. H., Hahn, K. J. and Barth, M. L., "Experimental Toxicology of Pyrolysis and Combustion Hazards," *Environ. Health Perspective 11*, pp. 191-196 (1975).
25. Lawrence, W. H., Raje, R. R., Sigh, A. R. and Autian, J., "Toxicity of Pyrolysis Products: Influence of Experimental Conditions," The MSTL/UT and NASA/JSC Procedures, *J. Combust. Toxicology*, 5, pp. 39-53 (1978).
26. Hilado, C. J., "Relative Toxicity of Pyrolysis Products of Polymeric Materials Using Various Test Conditions and Ranking Systems," *J. Consumer Product Flamm.*, 3, pp. 288-297 (1976).

BIOGRAPHY

Clayton Huggett

Clayton Huggett, Senior Scientist, Center for Fire Research, National Bureau of Standards, Gaithersburg, MD., received his B.S. degree in chemistry from the University of Wisconsin and his M.S. degree in organic chemistry and Ph.D. degree in physical chemistry from the University of Minnesota.

Dr. Huggett joined the National Bureau of Standards in 1970 where he was responsible for research on the ignition and combustion of materials, smoke and gas formation from burning materials, and fire retardants and fire extinguishing agents. In 1976, he became Chief of the Office of Extramural Research and in 1978, Deputy Director of the Center for Fire Research. In 1983, he resigned as Deputy Director in order to devote more time to technical problems. His current research interests include rate of heat release and toxicity of combustion products.

The Suitability of the DIN 53436 Test Apparatus for the Simulation of a Fire Risk Situation with Flaming Combustion

H. J. EINBRODT
RWTH Aachen, Abt. Hygiene

J. HUPFELD
Institut fur Schadenforschung Kiel

F. H. PRAGER
Bayer AG Leverkusen

H. SAND
BASF Ludwigshafen

SUMMARY

The fundamental considerations about the suitability of the DIN tube furnace for simulating the various fire models are discussed. By preliminary tests including the variation of test parameters it could be demonstrated, that in comparison to other decomposition models this method seems to be suitable to simulate real fire situations.

1. INTRODUCTION

THE TESTING AND evaluation of the toxicity of smoke requires a simulation of the fire exposure in a given risk situation. For economic reasons, testing has to be restricted to the main characteristic fire risk situations. For the subsequent animal exposure, a uniform composition of effluent products is required in order to obtain meaningful results.

The suitability of the DIN tube furnace for simulating·smoldering conditions has been proven by various investigations in different countries [1-8]. A lot of fundamental work with this piece of equipment was undertaken and is still going on in many laboratories [9-15].

With non flaming combustion, the countercurrent principle of furnace movement and air flow results in a more or less constant stream of smoke, but when spontaneous ignition occurs at higher test temperatures, greater deviations are observed.

At the same time, the speed of the flames may vary in relation to the movement of the furnace, depending on the characteristics of the material.

2. AIMS OF WORK

The aim of these investigations was to determine test parameters which ensure steady combustion under flaming conditions, consistent with the advance of the furnace, without necessitating any major changes to the existing test set-up according to DIN 53436. Thus it would be possible to evaluate relative acute toxicity of the gases evolved during burning with flaming combustion.

The test specimens may be ignited by

— spontaneous ignition or
— forced ignition (e.g. pilot flame, glow-wire or sparking).

Methods of ensuring steady burning of the test specimens include:

— splitting up the test specimens
— varying the speed of the furnace and
— varying the air supply.

To get ignition of the test specimen and/or the decomposition products, forced ignition by a pilot flame, a glow-wire or spark could be induced continuously or the temperature must be sufficiently high for self ignition and the concentration of gases must be adequate. In other words, an ignitable mixture must be present. With hydrocarbons, the lower ignition limit is at about 30-35 g/m^3 (1-5% by vol.).

At a decomposition speed (i.e. furnace speed) of 1 cm/min, the air flow selected (100, 200 or 300 l/h) then determines the minimum amount of sample tested as a function of the material properties. In the case of materials which only partially decompose or do not burn completely away, a correspondingly larger quantity must be used in order to obtain an ignitable mixture. The spontaneous ignition temperature is between about 250 °C (e.g. wood) and over 500 °C, depending on the constituents of the decomposition products as a function of exposure time. The temperature of the furnace must be set high enough to ensure ignition.

With polymeric hydrocarbons like, for example, polyethylene with a density of $\lambda = 1$ g/cm^3, specimens 1.5 cm wide and with thicknesses of approx. 0.3, approx. 0.6 and 1 mm respectively would be necessary for the above conditions.

In order to ensure constant flaming conditions on the test specimen, the rate of decomposition must correspond to the furnace speed. The energy required for steady flaming combustion must correspond to the quantity of heat produced externally (by furnace radiation, flame feedback). If the burning rate is too high, the decomposition induced with the aid of energy released during combustion will be faster than the speed at which the furnace moves.

3. TEST EQUIPMENT

The investigations were carried out on a test apparatus conforming to DIN 53436, Part 1 (Figure 1). The temperature of the furnace, measured by means of the reference element shown in Figure 2, can be adjusted over a range of temperatures up to approx. 750°C.

Figure 3 shows to what extent the thermal load on the specimen can be chosen and controlled by varying the temperature of the furnace. The heat flow intensity was measured by using a Topf radiometer [16] calibrated by means of a Medtherm (USA) type instrument as used in ISO TC 92/SC 1 "Reaction to Fire" [17].

In order to ensure continuous and more or less steady state combustion, the test specimen sections can be separated from one another according to Figure 4 by intermediate partitions made of inorganic materials like glass, mica or asbestos cement, a metal frame or simply by means of an air space (clearance was 5 mm in each case). Care has to be taken to ensure that this separation of the test specimens had no appreciable effect on the air flow profile, thus avoiding uncontrolled changes in the combustion process (e.g. soot formation).

Figure 1.

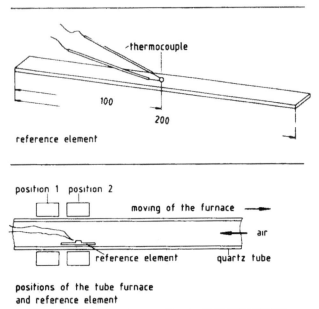

positions of the tube furnace
and reference element

Figure 2. Equipment according to DIN E53436/437.

In order to ensure continuous or more or less steady state combustion by forced ignition, an external ignition device is to be used. Construction details concerning an external ignition device by sparking are given in Figure 5 and 6. Detailed information concerning this modification of the DIN tube furnace is given in [18].

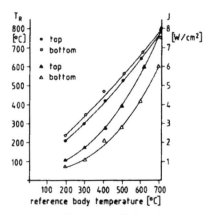

Figure 3a. Smoke temperature T_R and radiation intensity J as function of the reference body temperature.

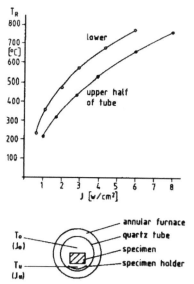

Figure 3b. Smoke temperature T_R versus radiation intensity J in the tube.

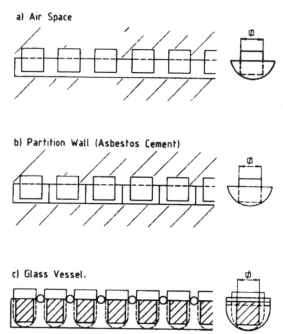

Figure 4. Partition of samples: a) air space, b) partition wall (asbestos cement), and c) glass vessel.

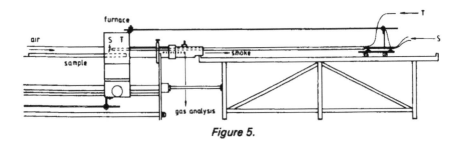

Figure 5.

In order to ensure continuous or more or less steady state combustion by variation of the rate of decomposition, technical limits of the DIN tube furnace procedure had to be overcome. To be able to vary the speed of the furnace to a broader extent, a modified tube length might be necessary.

4. TEST PROGRAM AND DISCUSSION OF RESULTS

The investigations were undertaken in three different laboratories. Each of them payed specific attention to one piece of the work. Spruce wood was used as the basis material of comparison.

Flexible polyurethane foam and expanded polystyrene with and without flame retardency were chosen to get some clarification to what extent the steady state burning conditions might be influenced by rapidly burning or melting materials.

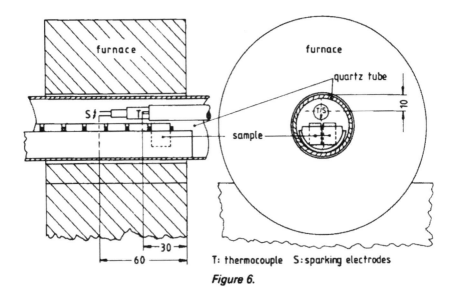

T: thermocouple   S: sparking electrodes

Figure 6.

4.1 The Effect of Air Throughput, Oxygen Concentration and Furnace Speed on Burning Behaviour

The influence of the furnace temperature and the air throughput on the thermal attack on the test specimens was determined in a series of preliminary test runs.

Figure 7 shows that increasing the amount of air (100, 200 and 300 l/h) while keeping the furnace temperature constant leads to a slight reduction (about 50 to 70 K) in the temperature of the thermocouples.

The investigations into the effects of furnace speed have shown that there are technical limits to this modification. If the exposure time of 30 minutes shall be retained, the speed of the furnace can only be increased by a factor of 2 (2 cm/min). With rapidly burning specimens this is not sufficient. By decreasing the oxygen concentration in the combustion air the burning velocity could be reduced down to 2 cm/min. This modification seems promising but needs further investigation.

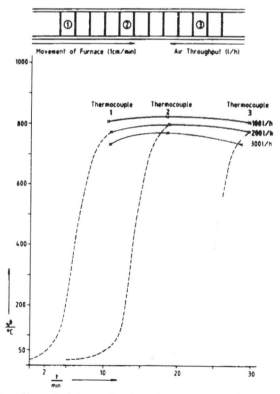

Figure 7. The effect of air throughput on the temperature level in DIN furnace.

4.2 Separation and Forced Ignition of Test Sample Sections on Burning Behaviour

Running the test series at a reference element temperature of 650 °C selfignition took place partly only. To get a steady state burning condition a forced external ignition was needed. By using the sparking device, which is moved parallel to the tube furnace, a continuous ignition of the test sample can be secured. After ignition took place, temperatures up to 900 °C were indicated by the thermocouple moving parallel with the sparking electrodes.

4.3 The Effect of Separation and Selfignition of Test Sample Sections on Burning Behaviour

The effect of separation of the specimens on the decomposition rate in the DIN furnace was studied at reference specimen temperature up to 700 °C.

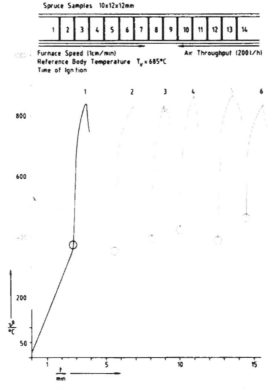

Figure 8. Influence of selfignition behaviour of samples to DIN tube furnace temperature level.

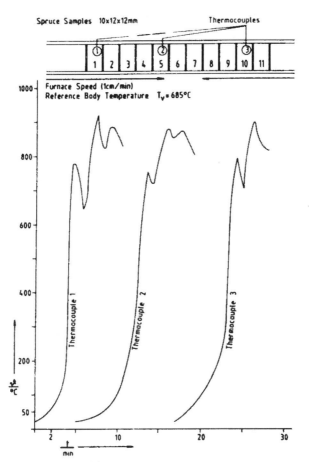

Figure 9. Influence of selfignition behaviour of samples to DIN tube furnace temperature level.

The results shown in Figure 8 (using spruce wood as an example) demonstrate how evenly the individual specimens ignite and burn (at a temperature of 700 °C) corresponding to the movement of the furnace.

The graph, Figure 8 clearly shows that, with spruce wood, for example, ignition can lead to an increase of between 400 and 450° K in the quartz tube. The increased thermal load resulting from spontaneous ignition of the spruce wood decomposition products is also shown in Figure 9.

Figure 10 shows that the temperature of an already burning specimen undergoes a further increase when the next specimen ignites. This is caused by the hot combustion products (see also Figure 9).

Figure 11 shows that concentrations of CO_2 and O_2 demonstrate

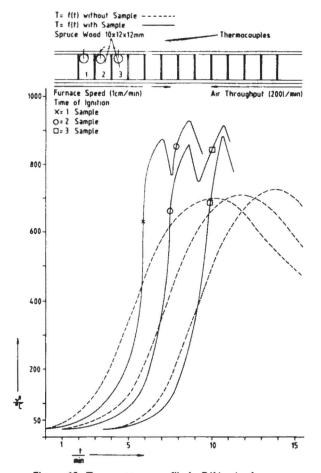

Figure 10. Temperature profile in DIN tube furnace.

fluctuations corresponding to the burning of the individual sample sections. Reducing the length of the sample sections should permit a better levelling off and flattening out of the peak values.

A specific arrangement of the test specimens as shown in Figure 12 should increase this levelling-off effect still further. Investigations are currently being prepared in this area.

5. FINAL CONCLUSIONS

The tests carried out on the DIN 53436-method to modify the procedure towards steady state flaming combustion conditions are only

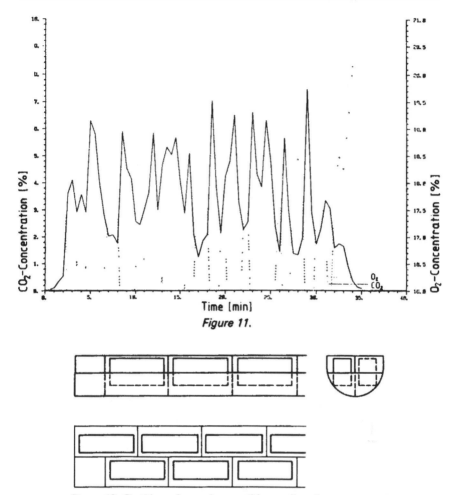

Figure 11.

Figure 12. Partition of samples; partition walls asbestos-cement.

preliminary. However, they have shown that in principle flaming conditions can be achieved even with rapidly burning materials.

The standardization of test parameters and the correlation of the results to real fire conditions require further investigation.

LITERATURE

1. Rumberg, E., and Reploh, H., "Analytische Untersuchung und medizinische Beurteilung von Brandgasen," Staal. Material-Prüfungsamt von Nordrh. Westf. Dortmd. (Okt. 1976).
2. Hofmann, H. Th. and Oettel, H., "Gesundheitsgefahren durch Ver-

schwelungsprodukte von organischem Material unter besonderer Berücksichtigung der Kunststoffe," *Z. VFDB, 3,* p. 79 (1968).

3. Kimmerle, G., "Methodology for Toxicological Analysis of Combustion Products," Polymer Series Conference, University Salt Lake City, Utah (1976).

4. Hoffmann, H. and Sand, E. H., "Weitere Untersuchungen zur relativen Toxizität von Kunststoff-Verschwelungsprodukten," *Kunststoff-Rundsch., 10,* pp. 413-416 (1974).

5. Kimmerle, G. and Prager, F. H., "Toxizität der Verschwelungs- und Verbrennungsprodukte verschiedener Chemiewerkstoffe und- fasern," *Kautschuk + Gummi, Kunststoffe, 33,* Nr. 5, p. 354 (1980).

6. Einbrodt, H. J. and Prjsnar, D., "Zur Toxizität gasförmiger Pyrolyse-Produkte aus Kunststoffen (Ein Beitrag zur Sicherheitstechnik)," *Wissenschaft und Umwelt, 2,* pp. 98-105 (1983).

7. Müller, W., "Tierexperimentelle Untersuchung zur Frage der Toxizität von Kunststoffverschwelungs-Produkten" *Diss. Uni Münster* (1976).

8. Boudene, C. and Jouany, J. M., "Extrude de la toxicote par voire deriene des produits de combustion et de pyrolyse des polyurethanes" (University of Paris), Unpublished information.

9. Baudene, C., Jouany, J. M. and Truhout, R., "Protective Effect of Water Against Toxicity of Pyrolysis and Combustion Products of Wood and Polyvinylchloride" (2nd Symposium on PVC, Lyon, July 1976) *J. of Makromol. Sci. Chem. All, 8,* pp. 1529-1545 (1919).

10. Klimisch, H. J., Hollander, H. W. and Thyssen, J., "Generation of Constant Concentrations of Thermal Decomposition Products in Inhalation Chambers. A Comparative Study with a Method According to DIN 53436 I. Measurement of Carbon Monoxide and Carbon Dioxide in Inhalation Chambers," *J. of Comb. Tox.,* Vol. 7, p. 243 (Nov. 1980).

11. Klimisch, H. J., Hollander, H. W. and Thyssen, J., "Comparative Measurements of the Toxicity to Laboratory Animals of Products of Thermal Decomposition Generated by the Method of DIN 53436," *J. of Comb. Tox., 7,* p. 209 (Nov. 1980).

12. Klimisch, H. J., "II. Measurement of Concentration of Total Volatile Organic Substances in Inhalation Chambers," *J. of Comb. Tox., 7,* p. 257 (Nov. 1980).

13. Herpol, C., "Biological Evaluation of the Toxicity of Products of Pyrolysis and Combustion of Materials," *Fire and Materials, 1,* pp. 29-35 (1976).

14. Szpilman, J., "Toxic Hazards During Fire. Insolation Materials," Dan Test Oct. 1982—Nat. Inst. to Testing Kopenhagen.

15. Szpilman, J., "Toxic Hazards During Fire. Wood Building Materials," Dan Test Aug. 1983—Nat. Inst. to Testing Kopenhagen.

16. Topf, P., ISO TC 92/SC1 WG4, Dok. 382, Instructions for 1980—Round Robin Test with ISO Smoke Box.

17. ISO TC 92/SC1 WG2, Dok. N54, Ignitability of Building Materials.

18. Institut für Schadenverhütung und Schadenforschung (IfS—Arbeits-Nr. 150 384053), Kiel.

Modeling of Toxicological Effects of Fire Gases: II. Mathematical Modeling of Intoxication of Rats by Carbon Monoxide and Hydrogen Cyanide

G. E. HARTZELL, D. N. PRIEST and W. G. SWITZER

Department of Fire Technology
Southwest Research Institute
P.O. Drawer 28510
San Antonio, Texas 78284
U.S.A.

ABSTRACT

Research in combustion toxicology over the past few years has led to a reasonable understanding and even quantification of some of the effects of fire effluent toxicants and, with the availability of a modest amount of both non-human primate and human exposure data, the combustion toxicologist is gaining increasing capability to assess and predict the toxicological effects of smoke inhalation.

This paper presents a mathematical approach, based on experimental data, for the prediction of both incapacitating and lethal effects on rats exposed to carbon monoxide and hydrogen cyanide. Elementary examples are given for computer simulation of the development of toxic hazards in fires and comparisons are made with actual experimental results. These comparisons show that computer-predicted times to toxicological effects lie within the standard deviation of experimental mean values.

The concepts described may ultimately obviate the routine use of laboratory animals for combustion toxicity testing and also open the way to the prediction of toxicological effects on humans exposed to fires.

INTRODUCTION

NUMEROUS HAZARDS TO humans result from exposure to the products of combustion of fires [1,2]. Among these are effects from heat, visual obscuration due to the density of smoke or to eye irritation, narcosis from inhalation of asphyxiants and irritation of the upper and/or lower respiratory tracts.

It is an ultimate objective to be able to model mathematically and, therefore, predict the toxicological effects of the inhalation of fire gases by humans. Scientific experiments under controlled laboratory conditions usually cannot be carried out using human subjects since the experiments themselves would be hazardous. Thus, research and development of methodology in this field has been largely of an analytical nature and/or limited to the use of experimental animals, usually rodents.

Two analytical approaches have been those reported at the National Research Council of Canada [3] and at McDonnell-Douglas under Federal Aviation Administration contract [4]. Both methods teach the summation of partial doses of toxicants in order to arrive at time-concentration conditions of toxicological consequence. These previously reported methods were lacking in satisfactory relevant toxicological data upon which to base the models and also in the assumption that the dose required to produce a given effect is constant under all time-concentration conditions. Furthermore, neither of the models has been tested with live subjects to enable the range of errors to be quantified.

Only after modeling methodology has been developed, based on relevant data for fire gas toxicants, and then validated, can extrapolation to human exposures be made. This, of course, is providing that both qualitative and quantitative differences in toxicological effects between laboratory animals and man are understood. Considerable progress is being made in developing this understanding such that models based on rodent studies are now expected to have relevant utility.

FIRE GAS TOXICITY

Although a wide variety of fire gases may be generated, the toxicant gases may be separated into three basic classes: the asphyxiants or narcosis-producing toxicants; the irritants, which may be sensory or pulmonary; and those toxicants exhibiting other and unusual specific toxicities.

In pharmacological terminology, a "narcotic" is defined as a drug which induces unconsciousness with loss of pain. In combustion toxicology, the term is used primarily in reference to asphyxiant toxicants that are capable of resulting in central nervous system depression with loss of consciousness and ultimate death. Although many asphyxiants may be produced by the combustion of materials, only carbon monoxide and hydrogen cyanide have been measured in fire effluents in sufficient concentrations to cause significant acute toxic effects.

The toxicity of carbon monoxide is primarily due to its affinity for the hemoglobin in blood. Even partial conversion of hemoglobin to carboxyhemoglobin (COHb) reduces the oxygen-transport capability of the blood (anemic hypoxia), thereby resulting in a decreased supply of oxygen to body tissues. Hydrogen cyanide, a very rapidly acting toxicant, does not combine appreciably with hemoglobin, but does bind with the trivalent ion of cytochrome oxidase in cellular mitochondria. The result is inhibition of the utilization of oxygen by cells (histotoxic hypoxia). Effects of the asphyxiant or narcosis-producing toxicants are dependent on the accumulated dose, i.e., both concentration and time of exposure, and increase in severity with increasing doses.

From an extensive review of methodologies for assessment of the incapacitating effects of the narcotic fire gases, both with rats and with non-human primates, it has been concluded that rats appear to be sensitive to approximately the same range of accumulated doses as may be deemed potentially hazardous to human subjects [5]. Thus, for both carbon monoxide and hydrogen cyanide, the rat is expected to be a reasonably appropriate model for the development of methodology for estimating toxicological effects on humans.

In view of the importance of the common asphyxiants, this study focused on the development of models for predicting the effects of carbon monoxide and hydrogen cyanide on rats, employing both incapacitation (leg flexion shock-avoidance) and within-exposure lethality as endpoints. Animal exposures, conducted in a typical 200-L National Bureau of Standards toxicity test chamber, used Sprague-Dawley rats positioned in tubular restrainers for head-only exposure. Atmospheres of carbon monoxide and hydrogen cyanide were generated from standard cylinder supply. Analyses for carbon monoxide were performed employing continuous non-dispersive infrared analyzers, with gas chromatography being used for hydrogen cyanide on an intermittent basis.

MATHEMATICAL MODELING

The basic concepts for toxicological modeling developed here involve the use of concentration-time relationships as a quantification of the "exposure dose" required to produce a given effect. From fundamental principles of toxicology, it has been demonstrated that concentration, mean exposure time to effect a given response, and percent of subjects responding are interrelated mathematically [6]. For example, a plot of mean exposure time required to effect a response as a function of concentration (Ct plot) also represents the EC_{50} (or LC_{50}) as a function of exposure time, or conversely, the ET_{50} (or LT_{50}) as a function of concentration.

Typical concentration-time relationships are illustrated in Figure 1 for incapacitation (loss of leg flexion shock-avoidance response) and lethality of rats from exposure to carbon monoxide. From this figure,

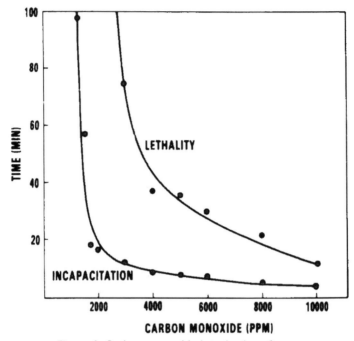

Figure 1. Carbon monoxide intoxication of rats.

the exposure dose required to cause incapacitation can, for example, be obtained by multiplying the values of any pair of coordinates; e.g., 2000 ppm × 20 min = 40,000 ppm-min exposure dose. The exposure dose required to cause an effect for a given toxicant is relatively constant (Haber's Rule) over the curved portion of the plot. For many purposes, the relative constancy of the Ct product can be used as a first approximation for estimation of the time-to-effect for a desired exposure concentration. However, for purposes of developing mathematical models having predictive value, further refinements must be made to take into account the fact that the Ct product exposure dose decreases with increasing concentration and may actually cover a two- to three-fold range.

Specific Ct exposure doses required to produce a given toxicological effect for a particular concentration may be determined as follows:

From the expression for Haber's Rule,

$$Ct = K, \tag{1}$$

simple rearrangement of terms produces

$$C = K\left(\frac{1}{t}\right). \qquad (2)$$

This is a linear equation, indicating that a plot of concentration against the reciprocal of time should be a straight line with the slope being K. Figures 2 and 3, using mean times-to-effect for exposure of rats to carbon monoxide, show that the relationship is, indeed, linear for both incapacitation and lethality. Similar linear relationships between concentration and the reciprocal of time-to-effect are shown in Figures 4 and 5 for incapacitation and death of rats due to exposure to hydrogen cyanide. (Summaries of laboratory data from which the plots in Figures 2 through 5 were constructed are shown in Tables 1 and 2).

Since the plots of $1/t$ against concentration do not pass through the origin, Equation (2) must be modified to include b, the y-axis intercept.

$$C = K\left(\frac{1}{t}\right) + b \qquad (3)$$

Equation (3) can be used directly for calculation of the EC_{50} or LC_{50} for any desired time of exposure once the constants have been determined. From Equation (3) can also be derived an expression for the Ct product exposure dose as a function of the concentration,

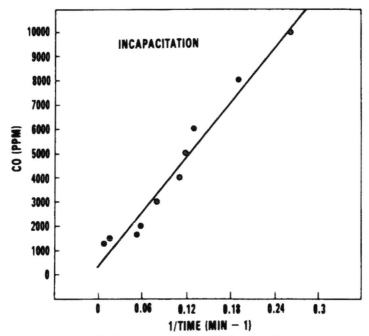

Figure 2. Carbon monoxide intoxication of rats.

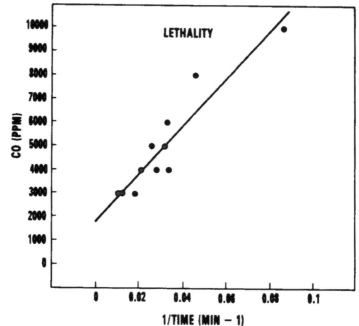

Figure 3. Carbon monoxide intoxication of rats.

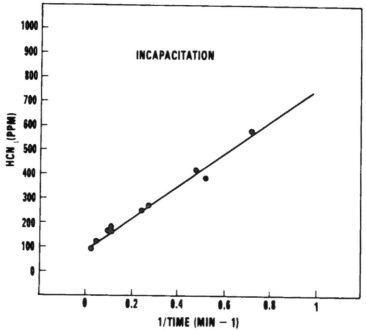

Figure 4. Hydrogen cyanide intoxication of rats.

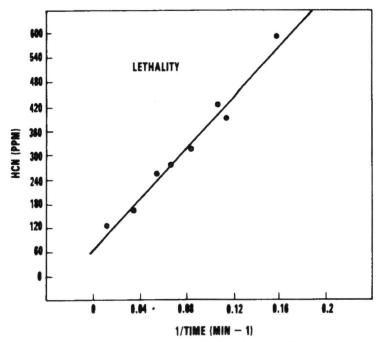

Figure 5. Hydrogen cyanide intoxication of rats.

Table 1. Carbon monoxide exposure data summary (N = 6 for each experiment).

CO Concentration (ppm)	Incapacitation (minutes)		Lethality (minutes)	
	Mean	S.D.	Mean	S.D.
1300	49.5	34.7		
1300	87.6	58.7		
1300	155.4	90.7		
1500	57.0	27.4		
1700	20.6	3.7		
1700	14.3	4.4		
1700	19.5	2.0		
2000	17.0	1.9		
2000	17.8	3.4		
2000	14.8	7.2		
3000	11.2	2.1	88.8	39.6
3000	11.8	3.6	79.9	42.3
3000	12.3	1.4	52.6	16.7
3000	13.2	4.6	78.1	36.0
4000	9.4	2.1	35.0	9.1
4000	9.1	1.8	46.5	21.6

(continued)

Table 1. (continued)

CO Concentration (ppm)	Incapacitation (minutes)		Lethality (minutes)	
	Mean	S.D.	Mean	S.D.
4000	8.0	0.6	29.4	5.2
5000	7.6	1.8	38.7	16.7
5000	7.3	0.7	37.1	8.0
5000	9.5	2.7	30.8	8.0
6000	7.4	0.8	29.9	10.4
8000	5.1	0.7	21.4	5.2
10,000	3.7	0.8	11.5	2.3

Table 2. Hydrogen cyanide exposure data summary (N = 6 for each experiment).

HCN Concentration (ppm)	Incapacitation (minutes)		Lethality (minutes)	
	Mean	S.D.	Mean	S.D.
95	43.8	10.3		
125	21.8	5.3		
127	21.0	4.6	83.3	32.1
165	8.8	3.0	*	*
165	10.9	3.9	28.2	3.8
190	8.9	2.5	*	*
253	4.1	1.2	18.0	5.5
275	3.7	0.9	14.6	1.9
314	1.8	0.7	11.6	0.7
392	1.9	1.1	8.7	0.7
425	2.1	0.6	9.2	1.2
592	1.4	0.5	6.3	0.6

*Not determined.

$$Ct = K \left(\frac{C}{C - b} \right), \tag{4}$$

where K and b are determined from the linear plot of experimental data. Equation (4) enables calculation of the Ct product exposure dose of the toxicant required to produce a given effect when present at any concentration. The slope K of the plot of concentration vs. $1/t$ represents the minimum value of the exposure dose required to effect a

50-percent response at sufficiently high concentrations such that the term $(C/C - b)$ approaches unity. Under these conditions, Haber's Rule (i.e., $Ct = K$) would be valid. The y-axis intercept, b, would thus appear to be a measure of deviation from Haber's Rule at relatively low concentrations of a toxicant.

In Table 3 are presented the values of K and b for both incapacitation and lethality for CO and HCN. Examples of calculated EC_{50}'s and $E(Ct)_{50}$'s for 15- and 30-minute exposures of rats are also shown. Data in Table 1 were derived from populations of 138 and 78 rats for incapacitation and lethality due to CO and populations of 70 and 46 rats for incapacitation and lethality due to HCN.

A restriction on the direct application of simple concentration-time relationships is that they are appropriate only for modeling constant concentration or "square wave" toxicant exposures such as those from which they were obtained experimentally. Fires do not produce constant toxicant concentrations, but rather, produce changing (usually increasing) concentrations. Máthematical models to predict effects from fire-generated toxicants must be able to accommodate varying concentrations which, in turn, involve varying Ct exposure doses required to effect incapacitation and/or death.

In Figure 6 is illustrated conceptually how this may be accomplished. From a plot of toxicant concentration as a function of time, incremental "doses" $(\overline{C} \times \Delta t)$ are calculated and related to the specific Ct "exposure dose" required to produce the given toxicological effect at the particular incremental concentration. Thus, a "fractional effective dose" (FED) is calculated for each small time interval. Continuous summation of these "fractional effective doses" is carried out and the time at which this sum becomes unity represents the time of greatest probability of occurrence of the toxicological effect in 50 percent of the exposed subjects. (As with all statistical data, there is a probability distribution for the predicted time-to-effect.)

The laboratory data Ct curve illustrated in Figure 6 is presented for

Table 3. Tabulation and use of modeling constants.

	K	b	EC_{50} 15-min (ppm)	EC_{50} 30-min (ppm)	$E(Ct)_{50}$ 15-min (ppm-min)	$E(Ct)_{50}$ 30-min (ppm-min)
			Incapacitation			
CO	36,509	233	2,667	1,450	40,004	43,499
HCN	698	92	139	115	2,078	3,458
			Lethality			
CO	102,874	1,778	8,636	5,207	129,544	156,214
HCN	3,130	66	275	170	4,120	5,110

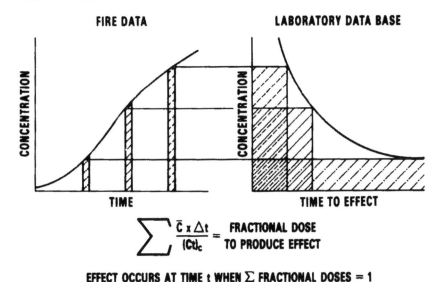

$$\sum_j \frac{\bar{C} \times \Delta t}{(Ct)_c} = \frac{\text{FRACTIONAL DOSE}}{\text{TO PRODUCE EFFECT}}$$

EFFECT OCCURS AT TIME t WHEN Σ FRACTIONAL DOSES = 1

Figure 6. Modeling of toxicological effects of fire gases.

conceptual purposes only. In practice, the specific Ct "exposure dose" corresponding to each incremental concentration is calculated from Equation (4).

The mathematical modeling concept illustrated in Figure 6 was tested for both incapacitation and lethality of rats exposed to increasing concentrations of carbon monoxide as might occur in a fire. Shown

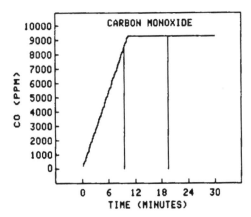

Time To Incapacitation = 10.0 Minutes
Time to Death = 20.8 Minutes

Figure 7. Test of FED model for carbon monoxide.

in Figure 7, the carbon monoxide concentration was ramped from 0 to about 9500 ppm over a period of about 10 minutes, followed by maintaining the concentration at the maximum level. By continuous summation of fractional effective doses for each 15 seconds of exposure, computer projections estimated incapacitation of exposed rats should occur at 10.0 minutes, with death expected at 20.8 minutes. Actual exposure of six rats to the ramped conditions yielded mean times of 10.2 ± 1.9 and 22.8 ± 3.5 minutes for incapacitation and death, respectively. The predicted times to both incapacitation and death were well within the standard deviations for the experimentally determined values.

Another test of the fractional effective dose model for carbon monoxide was conducted which employed a slower ramping profile for generation of the toxicant. The maximum CO concentration was attained at about 30 minutes. Results are shown in Figure 8. The computer-predicted time to incapacitation of 16.5 minutes was in excellent agreement with the experimental time of 17.2 ± 1.5 minutes determined using five rats. Death of the rats was predicted at 32.5 minutes. The experimentally determined mean time-to-death was 43.9 ± 13.9 minutes, with two of the five rats dead by 33.0 minutes.

The model was also tested for hydrogen cyanide, with exposure of six rats to HCN ramped from 0 to 195 ppm over about 12 minutes, followed by maintaining that concentration throughout the remainder of the exposure. Results are shown in Figure 9. Incapacitation and death were predicted by computer for 13.8 minutes and 32.2 minutes, respectively. Experimentally, mean times-to-incapacitation and death

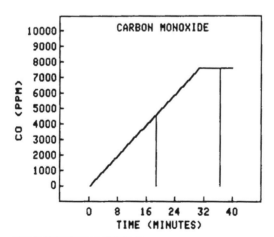

Time To Incapacitation = 16.5 Minutes
Time to Death = 32.5 Minutes

Figure 8. Test of FED model for carbon monoxide.

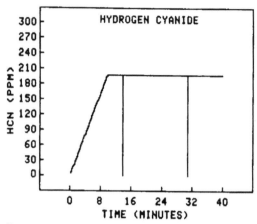

Figure 9. Test of FED model for hydrogen cyanide.

were 13.8 ± 2.3 minutes and 36.8 ± 11.9 minutes, respectively. Three of the six rats had died by 32 minutes.

A second test of the FED model for hydrogen cyanide was conducted with the concentration of HCN ramped from 0 to 270 ppm over 30 minutes. As illustrated in Figure 10, the model predicted incapacitation and death for 22.8 minutes and 34.0 minutes, respectively. Mean times-to-incapacitation and death of the six rats exposed

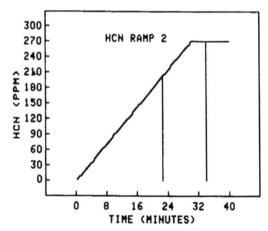

Figure 10. Test of FED model for hydrogen cyanide.

Table 4. Comparison of FED model with experimental data for carbon monoxide.

	Incapacitation (Minutes)		Lethality (Minutes)	
	Mean	Range (S.D.)	Mean	Range (S.D.)
FED Model	10.0	7.5-11.0	20.8	17.0-24.0
Experimental	10.2	8.3-12.1	22.8	19.3-26.3
FED Model	16.5	12.2-18.2	32.5	26.8-37.2
Experimental	17.2	15.7-18.7	43.9	30.0-57.8

in the ramped test were 20.8 ± 2.8 minutes and 34.2 ± 2.1 minutes, respectively.

A further refinement of the FED model enabled estimation of the error expected in the projection of times to incapacitation and death for CO and HCN. This involved determination of the K and b modeling constants for C vs. $1/t$ plots representing standard deviations of the original time to effect data. Analogous to the summation of fractional effective doses described for the projection of mean times-to-effect, summations were also carried out corresponding to Ct products representing standard deviations of the original laboratory data. A comparison of computer-projected and experimental data, including predicted errors and/or standard deviations is presented in Tables 4 and 5. It can be seen that predicted and experimental ranges for times to effect are sufficiently comparable as to suggest that model predictions are as accurate as experimental determinations.

The validity of the FED model has thus far been tested only for exposure of rats to single toxicants. However, with input data from appropriate laboratory experiments, the model would be expected to be quite appropriate for combinations of toxicants. The initial success of this methodology is anticipated to open the way to the combustion toxicity testing of materials without the necessity of using laboratory animals on a routine basis. Furthermore, the estimation of toxicological effects of smoke inhalation by humans exposed in a fire would also appear to be feasible.

Table 5. Comparison of FED model with experimental data for hydrogen cyanide.

	Incapacitation (Minutes)		Lethality (Minutes)	
	Mean	Range (S.D.)	Mean	Range (S.D.)
FED Model	13.8	12.0-15.8	32.2	25.2-35.2
Experimental	13.8	11.5-16.1	36.8	24.9-48.7
FED Model	22.8	21.0-24.2	34.0	30.5-37.0
Experimental	20.8	18.0-23.6	34.2	32.1-36.3

ACKNOWLEDGEMENT

This work was supported under U.S. National Bureau of Standards Grant No. NB83NADA4015 and under Southwest Research Institute Internal Research Project No. 01-9316.

REFERENCES

1. Kaplan, H. L., Grand, A. F. and Hartzell, G. E., *Combustion Toxicology: Principles and Test Methods*, Technomic Publishing Company, Incorporated, Lancaster, Pennsylvania, pp. 1-20 (1983).
2. Hartzell, G. E., Packham, S. C. and Switzer, W. G., "Toxic Products from Fires," *Am. Ind. Hyg. Assoc. J.*, 44 (4), pp. 248-255 (1983).
3. Tsuchiya, Y., "Dynamic Toxicity Factor—Evaluating Fire Gas Toxicity," *J. Combustion Toxicology*, Vol. 8, pp. 187-194 (August 1981).
4. Spieth, H. H. and Gaume, J. G., *A Combined Hazard Index Fire Test Methodology for Aircraft Cabin Materials*, Vol. 1, DOT/FAA/CT-82/36-1 (April 1982).
5. Kaplan, H. L. and Hartzell, G. E., "Modeling of Toxicological Effects of Fire Gases: I. Incapacitating Effects of Narcotic Fire Gases," *J. Fire Sciences*, Vol. 2, No. 4, pp. 286-305 (1984).
6. Packham, S. C. and Hartzell, G. E., "Fundamentals of Combustion Toxicology in Fire Hazard Assessment," *J. Testing and Evaluation*, Volume 9, No. 6, pp. 341-347 (November 1982).

Reporting Combustion Product Toxicity Test Results

CLAYTON HUGGETT

Senior Scientist
Center for Fire Research
National Bureau of Standards
Gaithersburg, MD 20899

TEST METHODS USED to study the inhalation toxicity of combustion products vary widely in experimental procedures and in methods of reporting results [1]. Consequently, it is very difficult to compare results obtained by different investigators on the same material or on different materials.

Recently I had occasion to examine the published data on poly (vinyl chloride) (PVC). Many references were found, differing in experimental conditions and expression of results. Various species of animals were used, furnace temperatures ranged from fixed to steeply rising, exposure chamber volumes varied from negligible to 1500 liters, gas flow rates ranged from zero to 20 liters per minute, exposure times varied from 12 to 240 minutes, and the experimental results were reported as percent mortality, LC_{50}, time to death, and critical temperature. On closer examination of the experimental details, however, it was found that many of these studies could be put on a common basis that permitted comparison.

The dose received by an experimental animal in an inhalation toxicology experiment, the actual amount taken up by the animal, is usually unknown. It can be assumed, however, that the actual dose is proportional to the exposure dose, the integral of the concentration of the combustion products over the exposure time. Methods of estimating the exposure dose for various experimental arrangements are given in [1]. The effective dose then is the exposure dose which produces the observed effect in the animal. If the observed effect is death the LED_{50} is the exposure dose which (statistically) causes death of 50 percent of the animals during the exposure and post-exposure observation period.

Ten references were found from which LED_{50} for PVC combustion products could be estimated [2–11]. Samples were not always adequately iden-

Year	Species	Heating Pattern	Exposure Time Min.	Exposure Chamber Vol. Liters	Exposure Flow Conditions l/min	Post Exposure Observation	Report	LED_{50} g min m-3	Reference
1969	Rat	3°C/min	140	7	2.83	7 days	mortality	450	2
1973	Rat	Rapid Rise	240	1500	static	7 days	LC_{50}	8640	3
1974	Rat	DIN 400°C	30	neg.	8.3	none	mortality	600	4
1974	Mouse	Fixed 350°C	15	56	4	7 days	mortality	~1600	5
1977	Rabbit	DIN 400°C	30	neg.	2	15 days	LC_{50}	1000	6
1977	Rat	Fixed 600°C	Time of death	12.6	recirculated	none	t_D	1200	7
1981	Mouse	20°C/min.	30	2.3	20	10 min.	LC_{50}	350	8
1982	Guinea Pig	50°C/min. to 500°C	30	neg.	1.1	2 hrs.	mortality	>700 <2000	9
1982	Rat	Fixed 575°C nonflaming 625°C flaming	30	200	static	14 days	LC_{50}	600 nonflaming 520 flaming	10
1983	Rat	DIN 500°C	30	neg.	3.3	none	mortality	<1860	11

tified but an attempt was made to include only rigid PVC which was assumed to be at least 90% PVC homopolymer. Experimental descriptions were sometimes incomplete and it was necessary to make plausible assumptions concerning operating procedures. In some cases, only upper and lower limits of toxicity could be estimated.

The results are summarized in the table. Despite the uncertainties and wide variations in procedure, the LED_{50} results are surprisingly consistent. Outliers are easily recognized and can be rationalized in terms of the peculiarities of the test method. Long exposure time or recirculation of the products leads to decay of the concentration and over estimation of the exposure, producing low toxicity values (high LED_{50}). Deaths during post exposure observation decreases the LED_{50}. In general, any lag in thermal decomposition or product decay during exposure in a static system will tend to decrease the observed toxic potency.

The lower values of the LED_{50} will be the most reliable. High values indicate possible deficiencies in the test method. The experimental procedure described in [8] gives a well defined exposure dose.

It appears that experimenters may enjoy considerable latitude in devising toxicity tests as long as a few simple precautions are observed and the results are expressed in exposure dose units.

I suggest that combustion toxicologists report, in addition to any other experimental observations, the exposure dose in their experiments. Such a practice would facilitate comparison of results between experiments and allow the extension of laboratory results to the conditions of variable product concentration and exposure time encountered in real fires.

REFERENCES

1. Huggett, C., *J. Fire Sci.* 2 (January/February 1984).
2. Cornish, H., Abar, A. L., *Arch. Environ. Health*, 19, p. 15 (1969).
3. Cornish, H., in *Symposium on Products of Combustion of (Plastic) Building Materials*, Armstrong Cork Co. (1973).
4. Hofmann, H. Th., Sand, H., *J. F. F/Comb. Tox.*, 1, p. 250 (1974).
5. Kishitani, K., Nakamura, K., *J. F. F/Comb. Tox.*, 1, p. 104 (1974).
6. Boudene, C., Jouany, J. M., Truhart, R., *J. Macromol. Sci-Chem.*, All, p. 1529 (1977).
7. Hilado, C. J., Crane, C. R., *J. Comb. Tox.*, 4, p. 56 (1977).
8. Alarie, Y., Anderson, R. C., *Tox. & Appl. Pharm.* 57, p. 181 (1981).
9. Jaeger, R. J., Skornik, W. A., Heimann, R., *Am. Ind. Hyg. Assoc. J.*, 43, p. 900 (1982).
10. Levin, B. C., Fowell, A. J., Birky, M. M., Paabo, M., Stolte, A., Malek, D., NBSIR 82-2532 (1982).
11. Herpol, C., *Fire & Mat.*, 7, p. 193 (1983).

Modeling of Toxicological Effects of Fire Gases: III. Quantification of Post-Exposure Lethality of Rats From Exposure to HCl Atmospheres

**G. E. HARTZELL, S. C. PACKHAM[1], A. F. GRAND
and W. G. SWITZER**

Department of Fire Technology
Southwest Research Institute
P.O. Drawer 28510
San Antonio, Texas 78284
U.S.A.

ABSTRACT

This paper, the third in a series of publications on the modeling of toxicological effects of fire gases, addresses the quantification of post-exposure lethality of rats from exposure to hydrogen chloride atmospheres. Experimental $L(Ct)_{50}$ values for HCl varied from about 80,000 ppm-min (5-minute exposure to 16,000 ppm) to about 170,000 ppm-min (60-minute exposure to 2800 ppm). Relevant data involving non-human primate exposures are cited, which suggest the comparability of the rat and the primate for the purpose of assessing lethal doses of HCl.

INTRODUCTION

ALTHOUGH A WIDE variety of fire gases may be generated, the toxicant gases may be separated into three basic classes: the asphyxiants or narcosis-producing toxicants; the irritants, which may be sensory or pulmonary; and those toxicants exhibiting other and unusual

[1]BETR Sciences, Incorporated, 996 South 1500 East, Salt Lake City, Utah, 84105.

specific toxicities. The roles of the major narcosis-producing toxicants, carbon monoxide and hydrogen cyanide, were discussed in previous papers of this series, with elementary mathematical models advanced for the assessment of toxic hazard development [1,2].

Irritant effects, produced in essentially all fire gas atmospheres, are normally considered by combustion toxicologists as being of two types. These are sensory irritation, including irritation both of the eyes and of the upper respiratory tract, and pulmonary irritation. Most irritants produce signs and symptoms characteristic of both.

Airborne irritants enter the upper respiratory tract, where nerve receptors are stimulated causing characteristic physiological responses, including burning sensations in the nose, mouth and throat, along with secretion of mucus. Sensory effects are primarily related to the concentration of the irritant and do not normally increase in severity as the exposure time is increased. There is no evidence that sensory irritation, *per se*, is physically incapacitating, either to rodents or to primates. On the contrary, recent studies involving exposure of baboons to hydrogen chloride have shown that even massive concentrations (up to 17,000 ppm for 5 minutes) are not physically incapacitating and do not impair escape [3].

Of potentially greater importance with both rodents and primates is that, following signs of initial sensory irritation, significant amounts of inhaled irritants are quickly taken into the lungs with the symptoms of pulmonary or lung irritation being exhibited. Tissue inflammation and damage, pulmonary edema and subsequent death often follow exposure to high concentrations, usually after 6 to 48 hours. Unlike sensory irritation, the effects of pulmonary irritation are related both to the concentration of the irritant and to the duration of the exposure.

Since studies using both non-human primates and rats have shown that neither incapacitation nor within-exposure death occurs from HCl up to exposure doses which result in post-exposure lethality [3], the described experiments with HCl atmospheres were directed toward quantification of post-exposure (up to 14 days) lethal doses.

EXPERIMENTAL

Animal exposures were conducted using the test apparatus shown in Figure 1. The exposure chamber was a typical 200-L National Bureau of Standards smoke toxicity test chamber, modified by incorporation of the animal isolation system illustrated in Figure 2. The animal isolation system was used to enable the preparation of test atmospheres with the animal unexposed, but in position for instantaneous "square-wave" exposures.

Prior to the generation of each HCl test atmosphere, adult male Sprague-Dawley rats were placed in tubular restrainers which were then inserted into the animal isolation system in preparation for head-

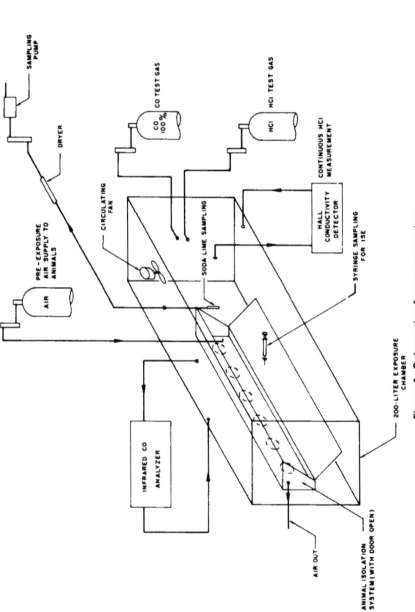

Figure 1. Schematic of exposure system.

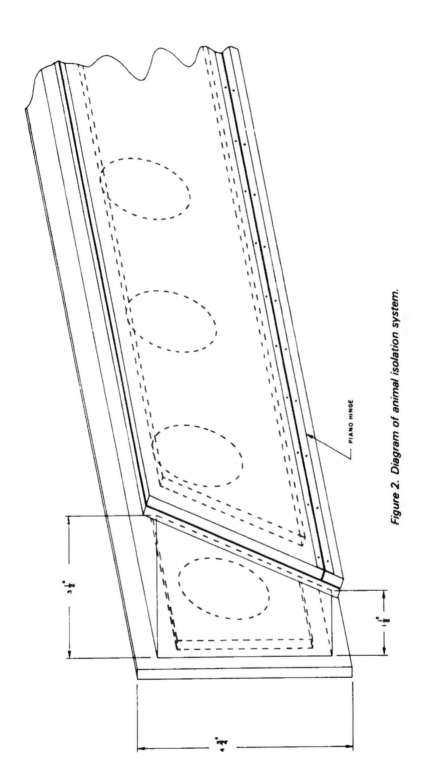

Figure 2. Diagram of animal isolation system.

only exposure of the subjects. A flow of air was supplied to the isolation system to purge CO_2 produced from normal animal respiration while preparation of the HCl atmosphere was in progress.

Generation and monitoring of hydrogen chloride (HCl) gas in the system was somewhat difficult due to the instability of HCl in the atmosphere. (Dry HCl reacts with moisture in the air, causing deposition onto surfaces and consequential loss from the atmosphere.) Furthermore, a truly quantitative continuous analyzer for HCl is not available. Dry HCl was metered into the system from a compressed gas cylinder through a Teflon® and glass flowmeter. The regulator and flowmeter were kept as dry as possible between test runs by purging them with nitrogen.

Sampling and analysis of HCl were performed by three different techniques, summarized as follows:

1) Batch sampling was accomplished using dry soda-lime adsorption tubes. Subsequent analysis of chloride ion in the aqueous extract from the soda-lime was performed by titration [4].
2) Intermittent gas sampling was done using 60-mL syringes containing 5-mL of absorbing solution. Direct (and immediate) analysis of chloride ion in solution was performed by ion-selective electrode (ISE).
3) Continuous sampling and analysis were performed using an atmosphere recirculating pump and a Hall Conductivity Detector. This detector was calibrated immediately prior to each run by technique No. 2.

Technique No. 1 (soda-lime tube) was relied upon for quantitative assessment of HCl, while the syringe method (No. 2 above) was used as a backup. The continuous analysis method was used to help establish and maintain a constant level of HCl during actual test runs. The latter technique was accomplished by continuously withdrawing and recirculating a portion of the atmosphere from the chamber, with a small volume (usually 1–4 cc/min) of the stream diverted for analysis. This was diluted with pure, dry air and directed concurrently with n-propyl alcohol through the conductance cell. The effluent solvent stream was not recirculated through the conductance meter. Calibration of the continuous monitor was checked during each test run. As HCl was introduced step-wise into the chamber (prior to exposure of the animals), a comparison was made between the mV output of the conductivity detector and an analysis of gaseous HCl in ppm using the syringe/ISE method.

During each actual exposure experiment, the HCl concentration was monitored with the continuous technique, with addition of HCl being made as necessary in order to maintain the desired level. Samples of the atmosphere in the exposure chamber were taken with soda-lime adsorption tubes over prescribed time intervals for the duration of each run. For example, soda-lime tube samples were taken over 5-minute durations for the 30-minute runs (6 samples), for 3 minutes during the 15-minute runs and for 1 minute during the 5-minute runs (5 samples). Syringe samples were also taken frequently during each run, for comparison with soda-lime tube analysis. All samples were taken at a height in the chamber approximately in line with the noses of the animals.

When the required HCl concentration was achieved and held constant, exposure experiments were initiated by terminating the air flow to the animals and releasing the door of the isolation system, thus instantly exposing the rats to the test atmosphere. In Figure 3 is shown HCl analytical data obtained during a typical animal exposure of 22.5 minutes duration.

RESULTS AND DISCUSSION

Post-exposure lethality LC_{50} values for HCl were determined for exposure times of 5, 10, 15, 22.5, 30 and 60 minutes. Data obtained for exposure of rats to at least five different concentrations of HCl for each exposure duration are summarized in Table 1. Also included in the table are statistically calculated LC_{50} and $L(Ct)_{50}$ values, along with the 95-percent confidence limits for each. All HCl concentrations are reported as ppm by volume.

Concentration-response plots for post-exposure lethality of rats are shown in Figures 4 through 9 and a plot of LC_{50} (post-exposure) as a function of exposure time is shown in Figure 10. (In contrast to analogous Ct plots for CO and HCN, which reflect mean times-to-effect within exposure for various concentrations, Figure 10 represents the time of exposure statistically estimated to cause 50-percent post-exposure lethality for any concentration of HCl.)

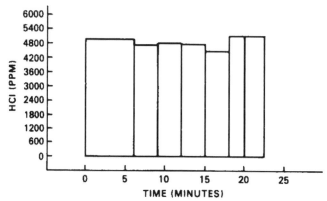

Figure 3. Typical analytical data for animal exposure to an HCl atmosphere.

Table 1. Hydrogen chloride exposure summary.

5-Minute Exposures				10-Minute Exposures				15-Minute Exposures		
Conc. (Avg. ppm)	Rats Exposed	14-Day Lethality		Conc. (Avg. ppm)	Rats Exposed	14-Day Lethality		Conc. (Avg. ppm)	Rats Exposed	14-Day Lethality
9,200	6	0		5,444	6	0		4,360	6	0
10,785	6	3		7,629	6	1		6,171	6	3
12,584	6	2		8,114	8	5		7,980	6	4
14,307	6	0		8,425	8	1		8,960	6	4
15,459	6	3		9,170	6	6		9,990	6	6
20,300	6	6								

LC_{50} = 15,900 ppm
95% Confd (11,540 21,890)
$L(Ct)_{50}$ = 79,500 ppm
95% Confd (57,700 109,400)

LC_{50} = 8,370 ppm
95% Confd (7,770-9,010)
$L(Ct)_{50}$ = 83,700 ppm
95% Confd (77,700-90,100)

LC_{50} = 6,920 ppm
95% Confd (5,380-8,900)
$L(Ct)_{50}$ = 103,800 ppm
95% Confd (50,700-133,500)

22.5-Minute Exposures				30-Minute Exposures				60-Minute Exposures		
Conc. (Avg. ppm)	Rats Exposed	14-Day Lethality		Conc. (Avg. ppm)	Rats Exposed	14-Day Lethality		Conc. (Avg. ppm)	Rats Exposed	14-Day Lethality
4,864	6	2		2,610	6	2		1,793	6	0
6,414	6	4		3,713	6	4		2,281	6	3
7,487	6	6		4,090	6	1		2,600	6	1
8,103	6	2		5,776	8	8		4,277	8	7
8,646	8	4		6,470	6	4		4,460	6	6
10,137	6	6		8,280	6	6		4,854	6	6

LC_{50} = 5,920 ppm
95% Confd (3,455 10,145)
$L(Ct)_{50}$ = 133,200 ppm
95% Confd (77,800-228,300)

LC_{50} = 3,715 ppm
95% Confd (2,540-5,435)
$L(Ct)_{50}$ = 111,450 ppm
95% Confd (76,200-163,000)

LC_{50} = 2,810 ppm
95% Confd (2,250-3,510)
$L(Ct)_{50}$ = 168,600 ppm
95% Confd (135,000-210,700)

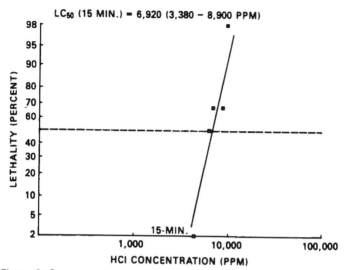

Figure 6. Concentration-response plot for HCl (15-minute exposure).

which subjects are exposed and is used here as an approximation of the actual dose which is, of course, unknown.) Experimental $L(Ct)_{50}$ values for HCl varied from about 80,000 ppm-min (5-minute exposure to approximately 16,000 ppm) to about 170,000 ppm-min (60-minute exposure to approximately 2800 ppm). This demonstrates that higher ex-

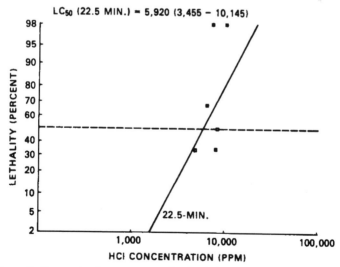

Figure 7. Concentration-response plot for HCl (22.5-minute exposure).

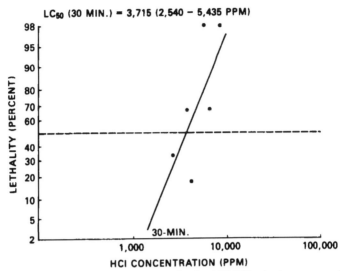

Figure 8. Concentration-response plot for HCl (30-minute exposure).

posure doses (Ct products) can be tolerated if the HCl concentration is relatively low. Conversely, progressively higher concentrations of HCl result in tolerance to decreasing exposure doses.

Post-exposure lethality data for gaseous HCl obtained in this study on rats are in good agreement with results reported previously for 30-minute exposures; however, data reported here for 5-minute ex-

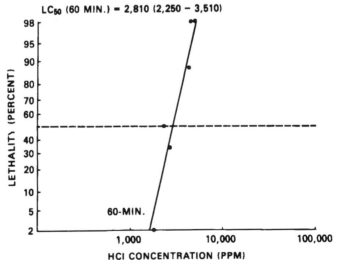

Figure 9. Concentration-response plot for HCl (60-minute exposure).

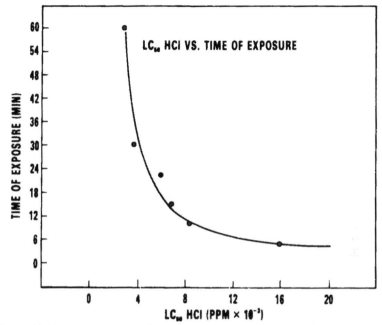

Figure 10. Post-exposure lethality of rats from exposure to hydrogen chloride.

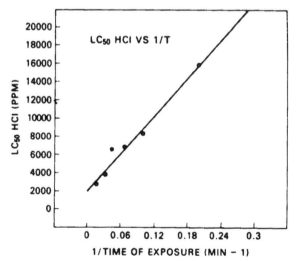

Figure 11. Post-exposure lethality of rats from exposure to hydrogen chloride.

posures appear to suggest somewhat greater toxicity for HCl [5]. Examination of the plot of HCl concentration versus exposure time for post-exposure lethality of rats in Figure 10 shows that concentration values for 5-minute exposures lie on an asymptotic portion of the curve. Thus, for short exposures, there would be no practical difference between concentrations above about 15,000-ppm HCl.

It is felt significant that the Ct exposure dose of HCl for 50-percent lethality (post-exposure) with rats lies in approximately the same range of exposure doses (150,000 ppm-min, based on 5- 15- and 30-minute exposure intervals) to which three non-human primates (juvenile baboons) were subjected without either incapacitation or post-exposure deaths occurring [6]. However, these exposures to non-human primates were quite severe and, in the opinion of the investigators, only slightly higher doses would likely have resulted in post-exposure deaths. Even the doses employed might have led to post-exposure lethality had medication (antibiotics and steroids) not been administered to the animals. From comparison of data between rat exposures and non-human primate experiments, it would appear that the rat is a reasonable model for primates as far as post-exposure lethal effects of HCl are concerned.

In view of the considerable differences in subject body weight, respiration rate and overall respiratory physiology between rats and primates, the comparable sensitivity with regard to pulmonary effects of irritants may appear to be rather surprising. The rat's lungs are somewhat protected, both by the large mucosal surface area of its nasal passages, which would be expected to absorb water-soluble HCl, and also by the rat's characteristic respiratory rate depression response to irritants. (Neither of these protective mechanisms are particularly effective for primates.) The extent to which the nasal scrubbing action of the rat reduces the concentration of HCl during inspiration is not known. Reduction of respiration rate and respiratory minute volume (RMV) due to HCl inhalation by rats has been quantified, however. For example, an HCl concentration of 1000 ppm reduces the rat's RMV to about 0.09 L/minute [7]. Man, with a normal RMV of about 6.40 L/minute [8], thus has about 70 times the RMV of the rat in the presence of HCl. However, with an estimated lung surface area of 70 m^2 [9], man has about 70 times the lung surface area expected for the rat. (The lung surface area of the guinea pig is reported to be about 1.1 m^2 [9].) Thus, in terms of actual mass of HCl per unit surface area of the lung, man and the rat could receive approximately equivalent doses from exposure to the same concentration of HCl in an atmosphere. This is, of course, an over-simplification which does not take into account other complicating factors. However, the comparison does offer some degree of explanation for the observation of the investigators that, for post-exposure lethal effects, the primate appears to be susceptible to approximately the same dose range of HCl as the rat. Sublethal pulmonary effects leading to respiratory dysfunction may be an entirely different matter, however.

ACKNOWLEDGEMENT

This work was supported under U.S. National Bureau of Standards Grant No. NB83NADA4015.

REFERENCES

1. Kaplan, H. L. and Hartzell, G. E., "Modeling of Toxicological Effects of Fire Gases: I. Incapacitating Effects of Narcotic Fire Gases," *J. Fire Sciences*, Vol. 2, No. 4, pp. 286-305 (1984).
2. Hartzell, G. E., Priest, D. N. and Switzer, W. G., "Modeling of Toxicological Effects of Fire Gases: II. Mathematical Modeling of Intoxication of Rats by Carbon Monoxide and Hydrogen Cyanide," *J. Fire Sciences*, Vol. 3, No. 2, pp. 115-128 (1985).
3. Kaplan, H. L., et al., "A Research Study of the Assessment Escape Impairment by Irritant Combustion Gases in Postcrash Aircraft Fires," Southwest Research Institute, Final Report, DOT/FAA/CT-84/16, U.S. Department of Transportation, Federal Aviation Administration, Atlantic City, New Jersey (1984).
4. Grand, A. F., Kaplan, H. L., Beitel, J. J., Switzer, W. G. and Hartzell, G. E., "An Evaluation of Toxic Hazards From Full-Scale Furnished Room Fire Studies," presentation at ASTM Symposium on Application of Fire Science to Fire Engineering, Denver, Colorado (June 26-27, 1984). ASTM STP 882, to be published 1985.
5. Darmer, K. I., Jr., Kinkead, E. R. and DiPasquale, L. C., "Acute Toxicity in Rats and Mice Exposed to Hydrogen Chloride Gas and Aerosols," *Amer. Ind. Hygiene Assoc. J.*, Vol. 35, pp. 623-631 (1974).
6. Kaplan, H. L., Southwest Research Institute, Unpublished Work.
7. Hartzell, G. E., Stacy, H. W., Switzer, W. G., Priest, D. N. and Packham, S. C., "Modeling of Toxicological Effects of Fire Gases: IV. Intoxication of Rats by Carbon Monoxide in the Presence of an Irritant," *J. Fire Sciences*, In Press (1985).
8. Phalen, R. F., *Inhalation Studies: Foundations and Techniques*, CRC Press, Incorporated, Boca Raton, Florida, p. 223 (1984).
9. *Mechanisms in Respiratory Toxicology*, Vol. 1, H. Witschi and P. Nettesheim, Ed., CRC Press, Incorporated, Boca Raton, Florida, p. 40 (1982).

The Toxicity of Fire
Effluents from Textiles
and Upholstery Materials

RAIJA KALLONEN

Technical Research Centre of Finland
Fire Technology Laboratory
Kivimiehentie 4
SF-02150 Espoo, Finland

ATTE VON WRIGHT[1] **and LEENA TIKKANEN**

Technical Research Centre of Finland
Food Research Laboratory
Biologinkuja 1
SF-02150 Espoo, Finland

KIRSTI KAUSTIA

Technical Research Centre of Finland
Chemical Laboratory
Biologinkuja 7
SF-02150 Espoo, Finland

ABSTRACT

The toxicities of the airborne combustion products of 11 textiles and 6 upholstery materials at medium temperatures (500–700 °C) were examined according to the standard DIN 53436 using rats as experimental animals. Also the chemical composition of the combustion products was determined. According to the results obtained there were some cases in which one or more individual components (mainly HCN and CO) of the airborne combustion products could be shown to be the main cause of toxicity.

In some cases no clear correlation between the CO concentration and carboxy-hemoglobin (COHb) levels could be observed. With polyurethane and flame retardant treated polyester fiber fill the COHb levels were surprisingly high

[1]Present address: VALIO Finnish Cooperative Dairies Association, P.O. Box 176, SF-00181 Helsinki.

despite the low CO amounts produced in the experimental conditions. The reverse was true with for example flame retardant treated cotton-viscose, which caused relatively low COHb levels despite of high production of CO. No fatal toxic effect could be linked with HCl despite some relatively high observed concentration.

INTRODUCTION

POISONOUS GASES AND smoke, rather than actual burns, are the main cause of death among the victims of fire accidents [1,2]. Therefore it is essential to determine the toxicity of the fire effluents of different materials used in furniture and decorations both in households and public buildings.

Kimmerle [3] and Herpol [4], among others, have studied the toxicity of the airborne combustion products of different materials and textiles in animal experiments. In these experiments the so called medium temperatures (400–700 °C) were found to be most critical with respect to toxicity. The toxicity of a certain material may also vary greatly in different pyrolysis temperatures.

The aim of this study was to investigate relationships between chemical composition and the toxicity of the airborne combustion products of certain materials. The materials studied included both textiles and upholstery materials. Both materials treated with flame retardant and untreated materials were used.

Toxicity was determined by exposing experimental animals (rats) to the airborne combustion products formed during the experiments.

MATERIALS

The materials were provided by different Finnish companies. They included textiles used in furniture, curtains, and bedclothing and synthetic foams and fiber fills used as upholstery in furniture and mattresses.

The textiles ·were cotton (Co), viscose (Vi), cotton-viscose blend (Co/Vi), wool, polyester (PES), modacrylic and polyvinylalcohol-polyvinylchloride (PVA-PVC). Several kinds of cotton were used: cotton as such, cotton made water-repellent with fluorochemicals and cotton treated with the following flame retardants (FR): N-methyloldimethyl-3-phosphonopropionamide (trademark Pyrovatex CP made by Ciba-Geigy), (preconcentrate of tetra-cis-hydroxymethylphosphoniumchloride and urea (trademark Proban made by Albright & Wilson) and a mixture of ammonium salts (trademark Flamentin made by Thor Chemie, West-Germany). Cotton-viscose was treated with Pyrovatex CP and viscose with 2,2'-oxybis(5,5-dimethyl-1,3,2-dioxaphosphorinane-2,2'-disulfide) (trademark Sandoflam made by Sandoz Colors & Chemicals). Polyester had been treated with some flame retardant not declared by the manufacturer. The flame retardants are subsequently referred in this paper using their trademark names.

The upholstery materials tested were FR-polyester fiber fill, neoprene foam and flexible polyurethane foams with the following trade names (made by Espe, Finland): E-35 (untreated), E-33 FR (treated with a flame retardant not declared by the manufacturer), E-35 PF (flame laminating) and HR-50 (high resilience).

METHODS

Combustion of the Samples

The test materials were decomposed in an airflow inside a quartz tube within a moving ring oven according to the standard DIN 53436 [5] (Figure 1).

The temperature within the tube was adjusted either to 500 °C or to 700 °C by calibrating the apparatus before the experiments using a reference element defined by the standard mentioned above.

The test samples were 40 cm long, about 1.5 cm wide and the total mass was adjusted to 0.12 g/cm.

The ring oven moved at constant speed (1.0 cm/min) along the test sample, which was kept within a cuvette inside the tube. Air was passed through the tube at a speed of 100 l/h. This airflow carried the smoke and fire gases into the mixing chamber (volume 1 l), into which more air was introduced at a speed of 300 l/h to dilute and cool the smoke-gas mixture. The temperature in the mixing chamber was measured using an ordinary thermometer. The total airflow was thus 400 l/h. The ratio of the mass of the test sample to the total air volume was 18 mg/l.

The test period was 30 minutes. Before the actual experiment the system was stabilized by leading the smoke and gases through a side tap to the exhaust.

The tube in which the combustion occurred was made of quartz the

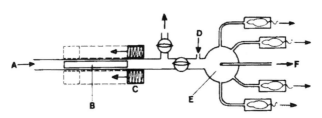

Figure 1. A schematic representation of the testing method for the toxicity of smoke and gases formed from the test material. (a) combustion air, (b) sample in a cuvette, (c) moving furnace, (d) dilution air, (e) distribution chamber and (f) gas sampling.

connecting tube of glass and the mixing chamber of acrylic plastic. The length of the connecting tube was less than 50 cm.

Chemical Analysis of Fire Gases

Gas samples for chemical analyses were taken from the mixing chamber. Co and CO_2 were monitored continuously using an infrared analyzing method (Hartmann & Braun AG, Model URAS 7 N). The gas sample was passed through the analyzer at a rate of 100 l/h and led into a paramagnetic O_2 analyzer (Hartmann & Braun Model Magnos 5-T). N_2 together with CO and CO_2 were used as calibration gases. N_2 (99.999% purity) was provided by AGA, Espoo, Finland. CO and CO_2 were used as N_2 mixtures. According to the manufacturers the CO and CO_2 concentrations were 455 ± 16 ppm CO + 4.29 ± 0.09% CO_2 (Alfax, Lidingö, Sweden) and 8420 ± 420 ppm CO + 4.77 ± 0,10% CO_2 (AGA, Lidingö, Sweden). CO-concentrations could be read directly as ppm (=μl/l), whereas CO_2 and O_2 concentrations were calculated as volume percentages. The results were expressed as average concentrations during the 30 min test period.

For the determination of HCN concentrations about 15 l of gas was pumped using suction through two sintered bubblers containing 100 ml water and 1 ml 10 M NaOH. The gas flow and volume were monitored using a rotameter and a gas volume meter, respectively. Disturbing sulphide ions were precipitated with lead nitrate solution and the reaction mixture was filtered. The concentration of CN^- ions was measured using ion-selective electrodes (CN^- electrode 94-06 with the reference electrode 90-02, both from Orion Research).

HCl concentration was measured by leading 15 l of gas through two sintered bubblers containing 100 ml 0.5 N sodium acetate.

The presence of disturbing sulphide ions was checked using nickel nitrate solution. The concentration of Cl^- ions was determined using the Cl^- sensitive electrode 94-17 and the reference electrode 90-02, both from Orion Research. The meter used in both HCN and HCl determination was an Orion Research microprocessor ionalyzer model 901. The results were expressed as average concentration during the test period. NO_2 was measured colorimetrically on the basis of the Griess-Saltzman reaction [6]. The gases were led from the mixing chamber through a glass fiber filter into the reagent mixture (sulfanilic acid/N-(1-naphtyl)-ethylenediamine), the adsorption of which was measured at 550 mm. (by a Spectronic 70 Spectrophotometer).

Volatile organic compounds were collected into an adsorbent (Tenax GC-60/80 or 2,6-diphenyl-p-phenylenedioxide) packed within a steel tube, through which the gas was passed at a speed of 0.5–1.0 l/min [7]. The internal diameter of the tube was 0.5 cm and the amount of adsorbent 0.35 g. The adsorbent was conditioned before the sampling in a nitrogen flow at 250 °C for 3 hours. Particular material was eliminated from the gas before the adsorption using a glass fiber filter.

The analysis of the volatile organic compounds was performed using a combined gas chromatograph-mass spectrometer (Jeol JMS-D100). The column was a Porapak Q/1 m steel column (program 60–250 °C, 10 °C/min). The sample was desorbed from the adsorbent tubes thermically at +170 °C and led into a cold trap (liquid nitrogen, −190 °C). From the cold trap the whole sample was fed into the gas chromatograph by warming it to 140 °C. The compounds were identified according to their retention times and mass spectra.

Toxicity Testing

Wistar rats (provided by Orion, Finland) were used as experimental animals. Before the experiments the rats were allowed to adapt for 7 days to the experimental animal unit of the Food Research Laboratory of the Technical Research Centre of Finland (VTT). Temperature was adjusted to +20 °C and the light and dark periods were 12 hours. Standardized mouse-rat fodder of Orion was given as food and tap water as drink. Before the experiments the medium weight of the rats was 172 ± 19 g.

Exposure of the rats to the combustion products of the test materials was performed at the Fire Research Laboratory of VTT. For each experiment 6 male and 6 female (or in some experiments 3 male and 3 female) rats were chosen. The variation of body weight within each test group was less than 20%. The diluted smoke-gas mixture was distributed from the mixing chamber into separate tubes each containing one rat (in principle the same as method III in Klimisch et al [8], see Figure 1). The exposure was for 30 minutes. Blood samples were taken either from the tips of the tails at the end of the exposure (living rats) or from the heart immediately after death from those rats which perished during the exposure. The carboxy-hemoglobin (COHb) content of the blood was measured spectrophotometrically (Perkin-Elmer 551S UV/VIS Spectrophotometer) within half an hour of sampling. For analysis the blood sample was haemolysed by ammonium solution. The absorbance of the sample was measured using wave lengths 575, 560 and 498 nm. The carboxy-hemoglobin content was calculated by the formula $(A_{575}-A_{560})/A_{498}$ and determined using the standard curve [9].

After the experiment the surviving rats were observed for two weeks at the experimental animal unit of the Food Research Laboratory. During this period they were weighed at intervals of 7 days. After the observation period the rats were sacrificed and the condition of the lungs and other internal organs was examined macroscopically.

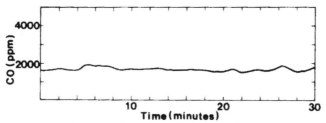

Figure 2. An example of an experiment with no flaming of the test material: CO content of the smoke gases of fluorochemical-treated cotton at 500°C.

RESULTS

Chemical Composition of the Airborne Combustion Products

The levels of CO, CO_2 and O_2 were almost constant throughout the exposure of the test animals in experiments with no flaming of the test material (Figure 2). In some cases shrinking or bending of the specimen at heat caused some variation. The standard deviation of the average CO concentrations calculated during five-minute time intervals (0–5, 5–10, 10–15, 15–20, 20–25 and 25–30 min) was in nonflaming tests between 2.8–31.3% with the average value of 11.8% and in flaming tests between 12.3–105.5% with the average value of 34.5%. If the test material ignited there were fluctuations in the concentrations of these three gases (Figure 3). When flames occurred the concentrations of CO and CO_2 usually increased, while that of O_2 decreased.

The average O_2 concentrations in the gas mixtures during the ex-

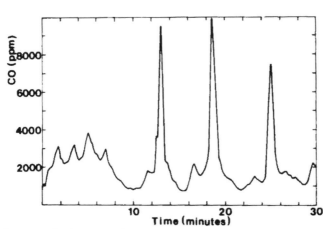

Figure 3. An example of an experiment with occasional ignition of the test material: CO content of the smoke gases of E 33-FR at 500°C.

periments were between 18 and 21% (Tables 1 and 2), providing enough oxygen for breathing (according to Herpol) [10], rats do not show signs of oxygen deprivation until the oxygen level is below 12-15%).

More CO_2 was generated from all the test materials at 700 °C than at 500 °C. The concentrations were not in themselves lethal, however, the highest average concentration measured being 2.0%.

Flame retardant cellulose fibers generated more CO than untreated cotton, especially at the higher decomposition temperature. No such effect was observed with cotton made water repellent with fluorochemicals. At both temperatures a concentration of CO considered lethal to rats during 30 minutes' exposure (LC_{50} = 4600-5000 ppm, lethal to 50% of rats in 30 min exposure [11]) was generated from Flamentin-treated cotton. At 700 °C lethal CO concentrations were also generated from Sandoflam treated viscose, Pyrovatex-CP treated cotton, Proban treated cotton and Pyrovatex-CP treated cotton-viscose blend. Upholstery materials did not generate notable amounts of CO (Tables 1 and 2).

At both temperatures relatively high concentrations of HCN (>100 ppm) were generated from modacrylic, wool, Flamentin treated cotton and FR-polyester fiber fill. These concentrations were lethal to rats (LC_{50} = 160 ppm [11]) except with FR-polyester fiber fill at 700 °C. Lower HCN concentrations were observed with textiles treated with Pyrovatex CP and Proban flame retardants. HCN generating materials contained nitrogen either in the base material, the flame retardant or in some other chemical added to the final product. The chemicals used as flame retardants or for some other purposes in FR polyester fiber fill and neoprene foam have not been declared by the manufacturer. The FR polyester fiber fill was probably mixed with nitrogen containing polyacrylonitrile, which is a commonly used binding agent in synthetic fiber fills.

Considerable concentrations of HCl (> 1000 ppm) were formed from the chlorine-containing materials PVA-PVC, modacrylic and neoprene foam at least at one of the test temperatures. However concentrations lethal to rats (LC_{50} = 3124 ppm [12]) were not detected.

Large amounts of sulphide estimated on the basis of PbS precipitate formation were present in the airborne combustion products of wool. This sulphide had to be eliminated before the ion-selective measurement of cyanide.

All the test materials generated oxygen containing organic compounds, eq. different aldehydes, ketones and alcohols. The most harmful among these were crotonaldehyde (LC_{50} = 4000 mg/m³ [12]), acetaldehyde (LCLo = 4000 ppm [12], lowest published lethal concentration to rats) and acrolein (LCLo = 8 ppm [12]). Most of the smoke gases contained aromatic compounds such as benzene, toluene and styrene.

In addition to the compounds mentioned above chloroacetaldehyde,

Table 1. Toxicity of fire effluents from fabrics and fillings at 700°C (18 mg/l.l.).

Material[5]	Flaming[1] +/-	CO[2] ppm	CO$_2$[2] %	O$_2$[2] %	HCN[2] ppm	HCl[2] ppm	Mortality % 30 min + 14 days			COHb%[4] at the end of the exposure	
							test 1	test 2	comb[2,3]	alive	dead
Fabrics											
Modacrylic	+	4200	1.1	21	1010	1590	100	–	100		41 (12)
Wool	---+	2500	0.7	19.5	610		100	100	100		32 (24)
Co FR Flamentin	–	6800	1.1	20	290	255	100	100	100		56 (6)
Vi FR Sandoflam	–	6400	0.7	20	<5		100	100	100		30 (12)
Co FR Proban	–	5400	1.4	19	50	15	67	100	83		44 (10)
Co/Vi FR Pyrovatex CP	–	4700	1.0	20	40	15	50	50	50	13 (2)	28 (3)
Co FR Pyrovatex CP	–	5700	1.1	20	30	20	67	17	38		74 (10)
PVA-PVC	+	4000	1.6	19	<5	630	17	0	8	20 (7)	64 (1)
PES FR	+	1600	1.9	19	<5	<5	0	17	8	22 (12)	
Co	+	2500	1.2	20	<5	<5	0	17	8	41 (22)	
Co Fluorochem. tr.	+	1900	1.1	20	15	<5	0	–	0	28 (12)	
Fillings											
PES FR fiber fill	–	2800	0.5	20	110	390	100	100	100		14 (12)
Neoprene foam	–	3400	1.4	19.5	35	760	83	67	75	18 (10)	21 (1)
HR-50 foam	+	1600	1.6	19	70		100	17	58	44 (5)	23 (7)
E-33 FR foam	+	1600	1.4	19.5	30		20	–	20	18 (10)	
E-35 PF foam	+	1400	2.0	18.5	55	35	0	0	0	9 (6)	
E-35 foam	+	1600	1.5	19.5	40	10	0	0	0	1 (4)	

[1] + flames, – no flames (all materials behaved similarly in the parallel tests except wool)
[2] Combined data of two tests except with Co, Co FR Pyrovatex CP, and wool combined data of three tests (the raw data combined with a new percentage calculated)
[3] Total number of rats 12 except with wool, Co, and Pyrovatex CP FR Co 24, and E-33 FR 10
[4] In parentheses number of rats from which blood sample could be taken
[5] The deviation of the combined data was in average 32% in mortality and the deviation of the mean 15% in CO, 14% in CO$_2$, 1% in O$_2$, 13% in HCN and 11% in HCl

Table 2. Toxicity of fire effluents from fabrics and fillings at 500°C (18 mg/l/l).

Material	Flaming[1] +/−	CO ppm	CO₂ %	O₂ %	HCN ppm	HCl ppm	Mortality %[2] 30 min + 14 days test 1	COHb%[3] at the end of exposure alive	COHb%[3] at the end of exposure dead
Fabrics									
Modacrylic	−	600	0.15	19.5	860	1370	100		18 (12)
Co FR Flamentin	−	6600	0.85		300		100		77 (12)
Co/Vi FR Pyrovatex CP	−	4000	0.7	20	45		42		75 (5)
Co	−	2400	0.85	20			42	40 (12)	
Co FR Proban	−	3100	0.8	20	50		33	25 (8)	1 (4)
CO FR Pyrovatex CP	−	3100	0.6	20	35		25	18 (9)	42 (3)
Co Fluorochem. tr.	−	1700	0.9	20	10	<5	25	38 (12)	
Vi FR Sandoflam	−	3200	0.35	20.5		2070	8	51 (12)	
PVA-PVC	+	1800	0.3	20.5			0	26 (12)	
PES FR	−	1000	0.2	20.5			0	42 (6)	
Wool	−	600	0.15	21	220		0	5 (12)	
Fillings									
PES FR fiber fill	−	2600	0.45	20.5	250	450	100		74 (12)
E-35 foam	+	2600	0.3	20	35		92	54 (9)	75 (3)
E-35 PF foam	+	1300	0.3	20.5	40		42	47 (12)	
E-33 FR foam	+	2000	0.55	20.1	25		40	9 (10)	
HR-50 foam	+	1500	0.55	20	30		17	52 (12)	
Neoprene foam	−	1000	0.6	20		1200	0	26 (12)	

[1] + flames, − no flames.
[2] Number of rats 12 except with E-33 FR 10.
[3] In parentheses number of rats from which blood sample could be taken.

dichloroethane (LCLo = 1000 ppm [12]), chloroprene (LCLo = 2280 ppm [12]) and chlorobenzene were formed from neoprene, PVA-PVC and FR-polyester fiber fill. Neoprene and PVA-PVC contain chlorine in their base material. FR-polyester fiber fill, on the other hand, had probably been treated with chlorine-containing chemicals, which is consistent with the observed formation of HCl.

Wool, modacrylic, FR-polyester fiber fill and neoprene gave rise to large amounts of different nitriles and cyano compounds, many of which, especially acrylonitrile (LCLo = 500 ppm), are poisonous. Wool and modacrylic contain nitrogen; FR-polyester fiber fill and neoprene were probably treated with nitrogen containing chemicals. At the higher temperature (700 °C) nitrogenous compounds were oxidized partly to nitrogen monoxide and -dioxide. The latter compound strongly irritates the respiratory tract. Hardly any nitrogen containing airborne combustion products were detected from the smoke of polyurethane. This is in agreement with a previous report [13] according to which the nitrogen compounds in polyurethane foam produce so called "yellow smoke," which needs a temperature exceeding 800 °C to form gas chromatographically analyzable compounds.

Sulphur compounds, e.g. sulphur dioxide, carbon disulfide and tiophenes, were detected in the smokes of wool and Sandoflam FR-viscose.

Representative examples of the contents of organic compounds in the smoke gases are given in Tables 3 and 4.

Biological Results

Most of the deaths among the experimental animals occurred during the 30 min exposure, especially at the temperature of 700 °C. At the temperature of 500 °C, delayed death also occurred during the 14-day observation period. The materials and the mortality of rats are presented in decreasing order of toxicity at both the experimental temperatures in Tables 1 and 2.

At 500 °C the most toxic materials (100% mortality) were modacrylic, Flamentin FR-cotton and PES FR fiber fill. At 700 °C modacrylic, wool, Flamentin FR-cotton, Sandoflam FR-viscose and FR polyester fiber fill had the strongest toxic effects, in decreasing order.

Some textiles were clearly more toxic at higher temperatures. This is demonstrated in Figure 4. The mortalities at 600 °C were obtained from our previous report [14], in which some of the test materials were the same as in the present study.

Very high (>65%) COHb-concentrations were found in the blood of the dead rats at the end of the exposure at 500 °C with Flamentin FR-cotton, Pyrovatex CP FR-cotton-viscose, FR-polyester fiber fill and E-35 foam and with Pyrovatex CP FR-cotton at 700 °C.

Smoke and gases formed at 700 °C from modacrylic, Flamentin FR-cotton and wool killed the experimental animals very rapidly, within 10 minutes after the start of exposure.

All the deaths during the two weeks' observation period occurred within the first week following the exposure. The delayed toxic effects were, in general, more pronounced with materials decomposed at 500 °C. Neoprene foam and Pyrovatex CP FR-cotton were exceptions, with delayed toxic effects appearing more clearly after decomposition at 700 °C.

Retarded weight gain was typical for the experimental animal

Table 3. Identified compounds in fire gases of wool and modacrylic.

Identified compounds	Wool		Modacrylic	
	500 °C	700 °C	500 °C	700 °C
2-methyl propene	1[1]	1	−	−
Sulfur dioxide	−	1	−	−
Acetonitrile	3	2	1	2
Furan	1	2	−	−
Butane	1	−	1	2
Carbon disulfide	1	1	−	−
Acrylonitrile	2	2	2	2
Dihydrofuran	1	−	−	−
2-propanenitrile	2	2	2	2
2-methyl propanal	2	2	−	−
Crotonaldehyde	1	−	−	−
2-butenenitrile	2	1	1	1
Methyl ethyl ketone	1	2	−	1
2-methyl-2-propenenitrile	2	2	2	2
2-methylpropanenitrile	1	2	2	2
Butenol	−	1	1	−
Benzene	2	2	2	2
Pyrrole	1	2	1	1
Pentadienenitrile	−	−	−	2
2-pentenenitrile	−	−	2	−
Pyridine	2	1	−	1
Isocyanobutane	−	2	−	−
Tolune	4	4	2	2
Chlorobenzene	−	−	−	1
Xylene	1	2	1	1
Styrene	1	2	1	−
Nitrogen dioxide	−	2	−	2

[1]The relative amount of a compound 1 = 0.1 − 1 μg/l
                                      2 = 1 − 10 μg/l
                                      3 = 10 − 100 μg l
                                      4 = > 100 μg/l

Table 4. Identified compounds in fire gases of FR polyester fiber fill and fabric.

Identified compounds	FR fiber fill		Fabric	
	500°C	700°C	500°C	700°C
Acetaldehyde	−	1[1]	3	2
Acetonitrile	1	1	1	1
Acrolein	1	2	2	−
Furan	2	3	2	2
Chloroethane	−	2	−	−
Chloracetaldehyde	1	−	−	−
Propyl aldehyde	3	2	2	2
Isopropyl alcohol	−	−	2	−
Acrylonitrile	2	2	−	−
Propanonitrile	2	2	−	−
2-methyl propenal	−	1	1	2
3-buten-2-one	1	−	2	2
Methyl ethyl ketone	1	2	2	2
2-methyl furan	−	−	2	1
Isothiocyanate ethane	−	−	2	−
1,2-dichloroethane	4	2	−	−
1,1-dichloroethane	4	2	−	−
Benzene	3	3	4	3
2-methyl pentene	−	1	−	−
Methyl pentanone	4	4	2	2
Toluene	2	2	3	2
Furfural	1	1	−	−
Chlorobenzene	−	2	−	−
Xylene	1	1	2	1
Styrene	−	1	2	−
Methyl pyridine	−	1	−	−
Dichlorobenzene	−	1	−	−
Benzoic acid[2]	−	−	4	4
Phtalic acid[2]	−	−	4	4
Nitrogen dioxide	2	2	−	−

[1] The relative amount of a compound 1 = 0 1 − 1 µg/l
          2 = 1 − 10 µg l
          3 = 10 − 100 µg/l
          4 = > 100 µg/l
[2] Analysis of the matter sublimated in large amounts on the walls of the sampling device

groups in which deaths occurred during the observation period. Signs of several kinds of lesions in the lungs (dark patches varying in size from "pinpoint" spots to areas covering whole lung lobes) were regularly observed in autopsies. Lung lesions were most marked in rats exposed to the pyrolysis products of PVA-PVC, although none of these rats died during the observation period, nor was there any sign of slowing down of the gain in body weight.

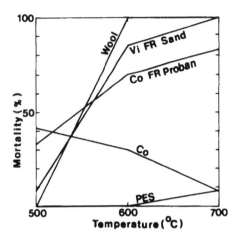

Figure 4. Toxicity of fabrics at different temperatures.

DISCUSSION

This study was started as a survey concerning the relationship between the chemical composition and animal toxicity of airborne combustion products. The results provided a mass of information but no basis for general conclusions. The most significant observations about both the experimental conditions and the results obtained are discussed below.

The variance between the results of the two series of experiments performed at 700 °C is given as a footnote to Table 1. The correspondence between mortalities was quite satisfactory for animal experiments. The variability of the concentrations of CO and CO_2 probably resulted from variation in the combustion process. This variation probably had an effect also on the repeatability of the animal experiments.

Measurement of the gas volume was found to be necessary because the gas flow was too uneven if only a rotameter was used, due to blocking of either the glassfiber filter of the hose or the sinter of the bubbler by particles in the smoke. Ion-selective measurement is relatively accurate, and so the variation in the concentrations of HCN and HCl probably resulted mainly from the burning process and sampling (these gases are readily soluble in water and very reactive).

The best known toxic compounds in fire effluents are CO, HCN and HCl. The importance of CO as a toxicant can be estimated on the basis of COHb concentration in blood. As mentioned in the previous sections remarkably high COHb concentrations were caused at 500 °C by Flamentin FR-cotton (77%) and Pyrovatex CP-FR-cotton viscose

(75%). At this temperature high COHb percentages were also found with FR-polyester fiber fill and E-35-polyurethane foam, although the CO concentrations in the smoke gases were not high (less than 3000 ppm). This at present unexplained phenomenon was also detected by Herpol [15] with certain materials.

At 700 °C high COHb concentrations were detected only with Pyrovatex CP FR-cotton (74%) and PVA-PVC (64%, though in one animal only). With some materials, e.g. Sandoflam FR-viscose (100% mortality) the COHb concentrations (30%) were low in spite of high concentrations of CO (6400 ppm) in the smoke gases. In these cases the other compounds present in the combustion products probably affected the absorption of CO into blood. E.g. irritants may reduce the volume of inhaled air and toxic combustion gases by affecting the breathing rate. Individual differences in COHb concentrations may also have resulted from the behaviour of the experimental animals (a nervous and restless animal breathes more heavily than a calm and quite one). As might be expected, dead animals had higher COHb concentrations than the survivors.

Only in some individual cases with Pyrovatex CP-FR-cotton-viscose and E-35 polyurethane foam at 500 °C together with Pyrovatex CP-FR-cotton at 700 °C, could CO conclusively be singled out as the major cause of death (high CO concentrations in combustion products together with high COHb contents in dead animals).

HCN was the obvious cause of death in tests with modacrylic at both the experimental temperatures and with wool at 700 °C. Both the high HCN concentrations detected and the very rapid death of the rats supported to this conclusion. However, no deaths were caused by wool at 500 °C despite high concentration of HCN (220 ppm) produced.

The combined effect of HCN and CO was probably the cause of the high toxicity of Flamentin FR-cotton at both 500 °C and 700 °C and also of modacrylic at 700 °C.

No fatal toxic effect could be linked with HCl despite some relatively high observed çoncentrations (e.g. PVA-PVC 2070 ppm). HCl might, however, have caused some of the lung lesions observed in the rats exposed to the combustion products of PVA-PVC and neoprene foam.

Relatively large amounts of hydrogen sulphide were generated from wool. Because the occurrence of this compound was not anticipated we had not taken any measures to analyse it quantitatively. As it is a poisonous gas, hydrogen sulphide might well have been an additional factor in the toxicity of this material.

Some of the upholstery materials (E-35, HR-50, neoprene and FR-polyster fiber fill) caused high mortality which could not be shown to result from the action of any single identified compound, although the considerable HCN concentration in the case of FR-polyester fiber fill might have been the main cause of the toxicity of this material.

Typical of the textiles treated with flame retardants was increasing toxicity at higher decomposition temperatures. On the other hand no such tendency could be discerned with upholstery materials treated with flame retardant. The toxicity of these materials at different temperatures depends solely on the material in question.

According to the results of these tests and others e.g. [15] toxicity of combustion products is not in connection with other combustion properties. E.g. modacrylic, wool and fire retarded cellulosic materials which produced and most toxic combustion products at 700 °C among the tested textiles and upholstery materials are relatively fire safe in respect to ignitability. Ignitability, rate of flame spread and other combustion properties in addition to toxicity of combustion products contribute to the toxic risk in fire and must also be taken into account.

ACKNOWLEDGEMENTS

The authors wish to thank Ms. Marja-Liisa Jalovaara, Ms. Aila Tuomolin and Ms. Tarja Vappula-Chirkow for excellent technical assistance. We also wish to thank Oy Finlayson Ab, Oy Tampella Ab, Valvilla Oy, Espe Oy and Martela Oy for supplying test materials.

REFERENCES

1. Harland, W. A. and Woolley, W. D., *Fire Int.*, *6*, 66, p. 37 (1979).
2. Birky, M. M., *Fire Mater.*, *3*, 4, p. 211 (1979).
3. Kimmerle, G. and Prager, F. K., *J. Comb. Toxicol.* 7, p. 42 (1980).
4. Herpol, C., *Fire Mater.*, *1*, 1, p. 29 (1976).
5. Producing Thermal Decomposition Products from Materials in an Air Stream and Their Toxicological Testing: Part 1. Test Method 53436. Deutsches Institut für Normung, Berlin (in German) (1981).
6. Standard Method Test for Nitrogen Dioxide Content of the Atmosphere: Test Method D-1607-69. American Society for Testing and Materials. Philadelphia (1970).
7. Pohjola, V., Manninen, A. and Häsänen, E., "Development and Study of Determination of Gaseous Impurities in Air," Technical Research Centre of Finland, Chemical Laboratory, Report 23 (in Finnish) (1979).
8. Klimisch, H.-J., et. al., *J. Comb. Toxicol*, 7, p. 243 (1980).
9. Hyvarinen, A., et. al. in *Clinical Laboratory Studies*, ed. by Werner Söderström, Porvoo, p. 450 (in Finnish) (1972).
10. Herpol, C. Minne, R. and Outryve, E. Van, *Comb. Sci Technol.* 12, p. 217 (1976).
11. Levin, B. C., et al., "An Acute Inhalation Toxicological Evaluation of Combustion Products from Fire Retarded and Non-Fire Retarded Flexible Polyurethane Foam and Polyester," Natural Bureau of Standards, NBSIR 83—2791, p. 18, Washington (1983).
12. National Institute for Occupational Safety and Health (NIOSH). Registry of Toxic Effects of Chemical Substances. U.S. Department of Health and Human Serivces, p. 761 (1980).
13. Woolley, W. D., *Br. Polym. J.*, 4, p. 27 (1972).

14. Kallonen, R., von Wright, A. and Tikkanen, L., "Toxicity of Fire Effluents from Textiles at Thermal Degradation Temperature of 600 °C," Technical Research Centre of Finland, Fire Technology Laboratory, Research Note 297 (in Swedish) (1984).
15. Herpol, C., *Fire Mater.*, 7, 4, p. 193 (1983).

Effect of Experimental Conditions on the Evolution of Combustion Products Using a Modified University of Pittsburgh Toxicity Test Apparatus

ARTHUR F. GRAND

Department of Fire Technology
Southwest Research Institute
P.O. Drawer 28510
San Antonio, Texas 78284

ABSTRACT

A study was performed on four materials, under several different combustion conditions, using a furnace and test parameters similar to one version of the "University of Pittsburgh Toxicity Test" procedure. Two of the materials were natural char-forming specimens, Douglas fir and cotton, and two were synthetic thermoplastics, polypropylene and polystyrene. Several different test conditions were used in an effort to determine the effect of varying the conditions from the "standard." Weight loss data, temperatures, and concentrations of carbon monoxide, carbon dioxide and oxygen were recorded.

A dramatic effect on sample combustion was observed with heating rate as the variable. Flaming combustion for three of the materials (all except cotton) occurred under the standard heating condition of 20° C/min. However, at a lower heating rate (10° C/min), the Douglas fir did not ignite, causing substantial differences in the evolution of carbon monoxide from that observed at the higher heating rate. The polypropylene ignited at both heating rates, while the polystyrene apparently ignited at the higher rate but generally not at the lower rate. The observations regarding carbon monoxide evolution as a function of heating rate help explain the lack of similarity between data from the Pittsburgh test and the NBS Toxicity Test.

Air flow rate was the other variable studied in this series of experiments. The quantities of gases evolved were not dependent on flow rate, as they should be if they agreed with theoretical calculations for a flow-through system. This observation led to the conclusion that there was substantial leakage of combustion gases from the furnace. It is recommended that these and other problems observed during the conduct of this program be considered before this procedure is adopted for regulatory purposes.

INTRODUCTION

THE PRODUCTS OF combustion of a material may be highly dependent on the combustion conditions. Thus, for many polymeric materials, the nature and concentration of evolved products may change with different temperatures, with nonflaming vs. flaming conditions, or with combustion in an oxygen-depleted environment compared to normal air. The toxicity of the resultant atmosphere is therefore also dependent on the nature of the combustion of the material.

The numerous combustion toxicity test methods in use have been reviewed recently and the advantages and drawbacks of each have been discussed [1]. One particular method, the University of Pittsburgh Toxicity Test (UPT), is of interest in this study. The apparatus and procedures originally developed for this test method at the University of Pittsburgh are described by Alarie and Anderson [2]. Anderson (now at Arthur D. Little, Inc.), et al., published a study for the State of New York [3] in which they also described the "University of Pittsburgh Method" and report data from their own laboratory. The development of the test method is also described and discussed in Reference [1] (pp. 83–108).

The objective of this program was to compare the amounts and rates of evolution of selected combustion gases (CO and CO_2) from four materials under several different combustion conditions. The purpose of the study was to see whether or not a change in conditions significantly affected the quantity of gas produced.

EXPERIMENTAL

The experimental setup used for these tests is shown schematically in Figure 1. The furnace, exhaust tube, arrangement of the weigh cell, etc., were intended to be as similar as possible to that of Anderson [4], which is a version of the same test apparatus and procedure as reported by Alarie and Anderson [2]. There apparently is no "standard" furnace. The one used by Alarie and Anderson at the University of Pittsburgh [2] is a "box" furnace measuring 25 cm × 23 cm × 34 cm (approximate volume of 19,500 cm³); whereas that used by Anderson at Arthur D. Little Co. is 34.3 cm (13½ in.) in diameter × 35.6 cm (14 in.) deep (about 32,900 cm³, although Dr. Anderson asserts in the New York State report that the volume is 42L, rather than 33L). The furnace used in this program is 30.5 cm (12 in.) in diameter × 35.6 cm (14 in.) deep (26,000

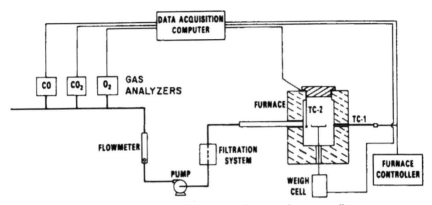

Figure 1. Schematic drawing of furnace and atmosphere sampling apparatus.

cm³ in volume); thus, it is very similar in internal dimensions to that used by Anderson.

The apparatus used in this study did not include an animal exposure chamber, nor was there provision for the diluting air in the airstream exiting the furnace. These were considered unessential to the present study and were therefore omitted. The effect of the diluting air can be accommodated by a simple calculation, as will be described later. The furnace was controlled by a thermocouple in the wall of the furnace, as specified by the manufacturer; whereas Alarie [5] apparently controls his furnace using a bare wire thermocouple in the airspace of the furnace. Data were recorded every 5 seconds for most experiments, every 10 seconds for some of the longer runs. An Apple IIe computer was used to log the data.

The procedure used in these experiments may be outlined as follows:

1) The gas analyzers were calibrated with compressed gas mixtures of known concentration. The load cell was calibrated with a known weight and was tared to zero with the sample pan in place.
2) The furnace was preheated, in most instances, to near 100 °C.
3) The air sampling pump was started and the flow adjusted to the desired flow rate.
4) The specimen was placed on the weigh cell and the run timer was begun immediately.
5) Data were automatically recorded over the duration of the run and consisted of time, sample weight, two thermocouple readings (bare wire in furnace airspace and furnace controller), and concentrations of carbon monoxide, carbon dioxide and oxygen.
6) The end of the run was usually determined by the concentration of the evolved gases (particularly CO) returning to baseline readings.

Table 1. Description of materials.

Cotton	— Cotton batting supplied by NBS.
Douglas fir	— Douglas fir board, 3/4-in. thick, supplied by Weyerhaeuser Company. Described as: clear vertical grain, trim grade.
Polypropylene	— Polypropylene resin (pellets) from Shell.
Polystyrene	— Polystyrene resin (granular) from Dow Chemical Company, Midland, Michigan; said to be a general purpose extrusion-grade resin. Density approximately 65 lb/ft^3.

Generally, but not always, the entire sample was consumed at that point (as monitored by the load cell).

7) Few additional observations were made during the run, since the combustion zone was not visible. However, ignition of the sample often resulted in explosive combustion.

The four materials tested were cotton, Douglas fir, polypropylene and polystyrene. They are described in Table 1. The quantities of materials used in these experiments were intended to be near the reported "LC$_{50}$" values for similar materials, as reported by Alarie and Anderson [2] or Anderson et al [3]. The one exception to this was Douglas fir, which was initially tested at the reported LC$_{50}$ of 50 g, but was reduced to 30 g due to the unexpectedly high evolution of carbon monoxide.

RESULTS

The test conditions utilized in this study are summarized in Table 2. The "standard" conditions used in the UPT procedures are 20 °C/min heat-up rate and 11 L/min gas flow rate out of the furnace. The other conditions used in this study included a lower rate of heating (10 °C/min) and lower and higher flow rates (approximately 6 and 18 L/min, respectively). In the UPT procedure, the 11 L/min is further diluted with 9 L/min of air to give a total flow of 20 L/min.

Table 2. Test conditions.

AIR FLOW	L/min		Heating Rate	
			10° C/min	20° C/min
		6	x	x
		11	x	Std. Cond.
		18		x

The bare wire thermocouple (TC-2 in the figure) was placed in the center of the furnace, near the specimen, in hopes of obtaining a better measure of the sample temperature than that provided by the shielded thermocouple (TC-1) in the wall of the furnace (which was connected to the furnace controller). The bare wire T/C consistently recorded temperatures 50° to 100 °C higher than the furnace controller thermocouple except under equilibrium conditions. In an effort to resolve the question of which temperature was the "real" temperature, the location of the control thermocouple was changed for some experiments towards the center of the furnace. When the controller T/C was nearer the center of the furnace, the difference in readings between the two T/C's was less. However, in this case, the control of the furnace heating rate was not as good. Since the controller/thermocouple arrangement was designed for maximum response, moving the T/C away from the wall of the furnace increased the time response and caused a less-constant heating rate. The controller T/C was finally moved back to the original location recommended by the manufacturer.

Data obtained in this study may be presented in several ways. Alarie and Anderson [2] plotted the weight loss and gas concentrations as functions of temperature of the furnace. Since they used the same rate of heating for all their experiments, one can easily overlay time on the x-axis instead of temperature. Those tests "began" at the point of 0.2% weight loss for any given material (determined ahead of time), which of course will be different for different materials. In this report, data are plotted as functions of both temperature, to compare with Alarie and Anderson, and time. For the plots reported herein, "zero" time corresponds approximately to 100 °C temperature at the controller thermocouple. All of the plots as functions of temperature also begin at 100 °C. Unless otherwise noted, the temperatures on the plots are that of the controller thermocouple (TC-1 in the figure). The flow rate for all of the test runs plotted was 11 L/min.

The maximum values of CO and CO_2 and minimum values of O_2 for each run are presented in Table 3. In Table 4, the quantity of CO evolved during each run is listed in terms of a concentration-time product (without any dilution or correction for relative flow out of the furnace), along with the ratio of CO_2 to CO. The data are presented in the table in terms of concentration-time because this product may be useful in predicting times-to-effect for CO as the toxic agent [6]. Examination of the data in these two tables provides some indication of test reproducibility as well as illustrating the extreme conditions that occur in the furnace during the burning of some of the materials. For example, several tests showed less than one percent oxygen and greater than 20% CO_2. Also, Table 4 will be useful for conclusions regarding the possible toxic effect of the atmospheres produced in these experiments (assuming that the primary toxic gas in these atmospheres is CO).

Table 3. Data summary of maximum or minimum gas concentrations.[a]

Run No.	Heating Rate, °C/min	Air Flow, L/min[b]	Maximum CO[c], ppm (Thousands)				Maximum CO$_2$, Percent				Minimum O$_2$, Percent			
			Cotton	Douglas Fir	Poly-propylene	Poly-styrene	Cotton	Douglas Fir	Poly-propylene	Poly-styrene	Cotton	Douglas Fir	Poly-propylene	Poly-styrene
1	20	6	—	32, 59[f]	6, 53	19	—	≥10[f]	10	9	—	4[f]	3	8
2	20	6	16, 16	39, 53[f]	9, 57	28	4	≥20[f]	19	16	17	4[f]	3	8
3	20	6	12[z]	53[d]	38[d]	8, 38[d]	4[z]	≥20[d]	18[d]	13[d]	18[z]	7[d]	3[d]	8[d]
4	20	6	19	—	—	—	6	—	—	—	15	—	—	—
1	10	6	15	43	11	7	4	6	14	1	17	14	7	19
2	10	6	14	—	7, 8	—	4	—	11	—	17	—	8	—
3	10	6	—	—	—	20	—	—	—	2	—	—	—	16
4	10	6	—	37	—	9	—	5	—	2	—	15	—	19
1	20	11	17	33, 52	6, 45	25, 44	5	≥20	≥20	11	16	7	<1	9
2	20	11	21	47, 15	5, 28[e]	44, 44	5	20	≥20[e]	16	16	9	1[e]	7
3	20	11	—	—	7, 100	—	—	—	17	—	—	—	<1	—
4	20	11	—	—	76	—	—	—	6	—	—	—	8	—
1	10	11	10	54	—	26	4	6	—	13	18	14	—	10
2	10	11	13	55	—	10	4	6	10	2	18	13	9	18
3	10	11	—	—	8, 8	9	—	—	17	1	—	—	5	19
4	10	11	—	—	9, 16	—	—	—	9	—	—	—	10	—
1	20	18	14	25	5, 34	14, 14	4	20	≥20	11	18	9	2	11
2	20	18	19	32, 17	5, 84	13	5	16	20	14	17	10	2	9
3	20	18	—	—	7, 91	—	—	—	18	—	—	—	3	—

[a] Sample weights were cotton—15 g. Douglas fir—30 g. polypropylene—10 g. polystyrene—6 g. except: [f]Douglas fir Nos 1-20-6 and 2-20-6—50 g and [z]cotton No 3-20-6—10 g.
[b] Actual flow rates were 5 7 L/min, 10 9 L/min and 16.5 L/min, respectively, for 6, 11 and 18.
[c] Two numbers indicate two peaks in CO concentration.
[d] Control thermocouple (TC1) repositioned to center of furnace.
[e] Furnace lid partially open.

Table 4. Concentration-time products for test runs.

Run No.	Heating Rate °C/min	Air Flow L/min[a]	Cotton			Douglas Fir			Polypropylene			Polystyrene		
			CO ppm-min (1000's)	CO_2 %-min	CO_2/CO[b]	CO ppm-min (1000's)	CO_2 %-min	CO_2/CO[b]	CO ppm-min (1000's)	CO_2 %-min	CO_2/CO[b]	CO ppm-min (1000's)	CO_2 %-min	CO_2/CO[b]
1	20	6	—	—	—	253.0[c]	133.7*	5.3*	133.7	66.0	4.9	78.3	26.9	3.4
2	20	6	126.0	73.2	5.8	217.6[c]	237.2*	10.9*	174.7	97.8	5.6	83.3	37.4	4.5
3	20	6	116.0[d]	60.0	5.2	141.2	168.2*	11.9*	134.4	99.8	7.4	85.9	35.6	4.1
4	20	6	152.1	111.1	7.3	—	—	—	—	—	—	—	—	—
1	10	6	158.4	110.4	7.0	474.8	139.5	2.9	205.9	132.2	6.4	75.6	14.4	1.9
2	10	6	145.5	86.4	5.9	—	—	—	95.3	111.5	11.7	—	—	—
3	10	6	—	—	—	—	—	—	—	—	—	103.7	15.9	1.5
4	10	6	—	—	—	326.8	103.6	3.2	—	—	—	77.9	20.3	2.6
1	20	11	120.6	87.1	7.2	132.9	176.5*	13.3*	133.1	86.9*	6.5*	89.3	30.7	3.4
2	20	11	117.8	89.5	7.6	107.3	175.4	16.4	75.7	66.5*	8.7*	96.7	39.9	4.1
3	20	11	—	—	—	—	—	—	289.7	88.9	3.1	—	—	—
4	20	11	—	—	—	—	—	—	268.2	55.0	2.1	—	—	—
1	10	11	124.0	83.5	6.7	457.5	109.0	2.4	—	—	—	88.1	32.7	3.7
2	10	11	142.7	96.1	6.7	463.8	132.3	2.9	—	—	—	73.1	20.3	2.8
3	10	11	—	—	—	—	—	—	65.4	105.9	16.2	85.2	22.2	2.6
4	10	11	—	—	—	—	—	—	132.0	79.9	6.1	—	—	—
1	20	18	92.2	71.4	7.7	45.2	151.6	33.5	103.1	73.8*	7.2*	46.8	22.5	4.8
2	20	18	97.8	73.8	7.6	81.8	118.0	14.4	137.2	79.6	5.8	43.9	28.4	6.5
3	20	18	—	—	—	—	—	—	191.1	83.5	4.4	—	—	—

* CO_2 reached or exceeded the limit of the gas analyzer; therefore, CO_2, sum and CO_2/CO ratio may not be accurate.

[a] Actual flow rates were 5.7 L/min, 10.9 L/min and 16.5 L/min, respectively for 6, 11 and 18.

[b] $CO_2/CO = CO_2$ (%-min) × 10 ÷ CO (1000 ppm-min).

[c] 50 g sample (other Douglas fir specimens were 30 g).

[d] 10 g sample (other cotton specimens were 15 g).

Sample weights (except as noted): cotton—15 g, Douglas fir—30 g, polypropylene—10 g and polystyrene—6 g.

DISCUSSION

Heating Rate

The two heating rates examined were 20° C/min and 10 °C/min. Typical weight-loss curves for Douglas fir vs. time and temperature are shown in Figures 2 and 2a, respectively. The patterns of weight loss vs. time for the two heating rates are not unusual, with a more gradual loss in weight for the slower heating rate. The weight loss curves as functions of temperature reveal that the slower heating rate actually caused degradation of the sample at a lower temperature. This may be due to the delay in heating the sample mass at the higher heating rate, which would not be surprising; however, the relationship of weight loss to temperature may be a critical factor in determining whether or not a sample ignites at a given heating rate.

Patterns of carbon monoxide concentration vs. time, also for Douglas fir, are shown in Figure 3. In these plots, substantial differences in CO evolution between the two heating rates are evident (the flow rate for all the plots was 11 L/min). The CO evolution at 20° C/min produced very sharp spikes in concentration over a shorter time duration than at 10° C/min. The shorter time duration for the higher heating rate is expected since it takes approximately twice as long for a given sample to thermally degrade at 10° C/min heating rate than at 20° C/min. However, it is surprising that substantially more CO is produced from the decomposition at 10° C/min. In the case of the Douglas fir, it is apparent that the same quantity of CO is not being produced under both heating conditions. The corresponding CO curves for Douglas fir are plotted as functions of temperature, rather than time, in Figure 3a. These do not show the differences described above because of the compressed scale for the 10° C/min run.

Data for oxygen consumption for the same Douglas fir samples are shown in Figure 4. The decomposition of the specimen at 20° C/min consumed substantially more oxygen than that at 10° C/min. Since the higher consumption of oxygen corresponded to a lower evolution of CO, it seems likely that the mode of decomposition at 20° C/min and at 10° C/min was different. The differences in combustion are apparent from the temperature plots in Figure 5. In this figure, the temperature of the bare-wire thermocouple in the air space (TC-2) is plotted versus the temperature indication of the shielded thermocouple in the wall of the furnace (TC-1). The TC-2 temperature curve for the 20° C/min rate shows repeated sharp spikes indicating flaming combustion; whereas at 10° C/min, the temperature shows only smooth undulations suggesting nonflaming combustion.

Further evidence of flaming versus nonflaming combustion for Douglas fir appears in Table 4. The carbon monoxide concentration-time products at 10° C/min were substantially higher than at the 20° C/min

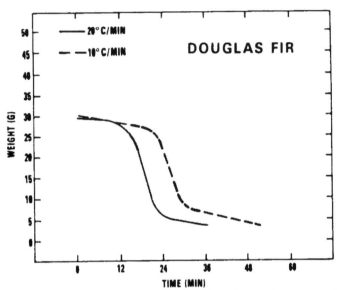

Figure 2. Weight loss of Douglas fir as a function of time (run nos. 1-20-11 and 1-10-11).

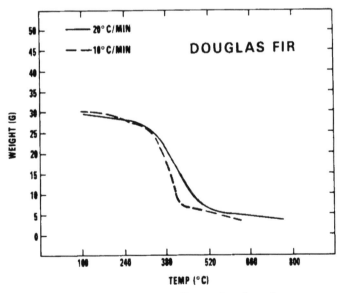

Figure 2a. Weight loss of Douglas fir as a function of temperature.

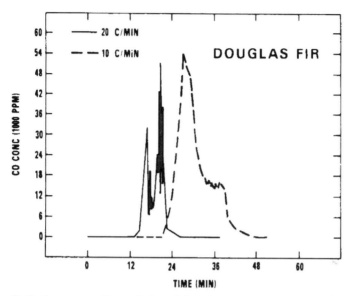

Figure 3. Carbon monoxide evolution of Douglas fir as a function of time (run nos. 1-20-11 and 1-10-11).

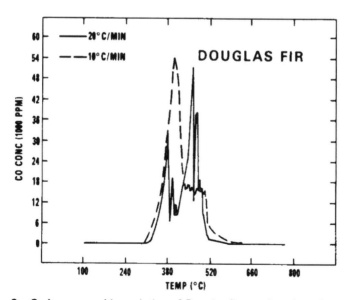

Figure 3a. Carbon monoxide evolution of Douglas fir as a function of temperature.

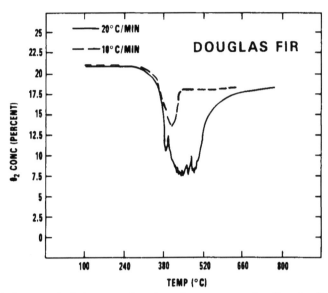

Figure 4. Oxygen depletion as a function of temperature for Douglas fir (run nos. 1-20-11 and 1-10-11).

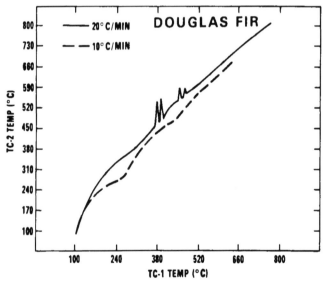

Figure 5. Temperature of a bare-wire thermocouple (TC-2) vs. temperature of the shielded controller thermocouple (TC-1) for Douglas fir at two different heating rates (run nos. 1-20-11 and 1-10-11).

heating rate (the average of the two runs at 10° C/min and 11 L/min was 461,000 ppm-min CO, while those at 20° C/min and 11 L/min averaged 120,000 ppm-min). Also, the CO_2 levels were lower for the 10°C/min run (approximately 121 %-min for 10° C/min and 175 %-min for 20°C/min). This caused the ratio of CO_2 to CO (also given in the table) to increase from less then 3 for the 10° C/min case to around 16 for 20° C/min. The conclusion that flaming combustion occurred during the higher heating rate and nonflaming combustion at the lower rate is consistent with these observations and with the shapes of the carbon monoxide evolution and oxygen consumption curves for the two heating rates. It is not clear why the weight loss curves at the two different heating rates are so similar for different combustion conditions.

The weight loss curves for cotton are presented in Figures 6 (vs. time) and 6a (vs. temp). As with the Douglas fir, the appearance of the curves as functions of temperature shows that the slower heating rate caused degradation at a lower temperature. The CO evolution from cotton is illustrated in Figures 7 and 7a. The maximum concentration of CO produced by the lower heating rate is approximately half that of the higher heating rate, but the CO evolution extends for approximately twice as long; therefore, the area under the concentration-time curve for the two heating rates is nearly the same (see Table 4). These data suggest that the same mode of combustion was occurring under the different heating rates and that the different CO curves are due only to the extended time of decomposition (it is interesting to note that the curves of CO vs. temperature in Figure 7a, as Alarie and Anderson would plot the data, do not give any indication that the total amount of CO produced was the same for the two test conditions). The cotton apparently was smoldering under both of the heating rates, as indicated by the lack of temperature spikes in the output of the TC-2 thermocouple (curves not shown).

Polypropylene and polystyrene are thermoplastic materials which melt readily under the influence of heat and do not char (as is the case with Douglas fir and cotton). The weight loss curves for polypropylene (Figures 8 and 8a) for the two heating rates are similar to those previously presented for the other materials, except for the more rapid loss in weight of the polypropylene. The extremely brief time during which weight was lost at 20° C/min heating rate is apparent in Figure 8.

Carbon monoxide evolution from polypropylene is shown in Figure 9 for the two heating rates. Although the data are plotted as functions of temperature, rather than time, it is apparent that there is a large discrepancy in the CO evolution from specimens subjected to the different heating rates. In these examples, much more CO was evolved at 20° C/min than at 10° C/min; however, it will be shown that both tests represent flaming combustion conditions. The corresponding oxygen consumption data for these two runs are plotted in Figure 10. Combustion at the higher heating rate, which produced the higher CO level, caused oxygen levels to decrease to nearly zero, while the consumption

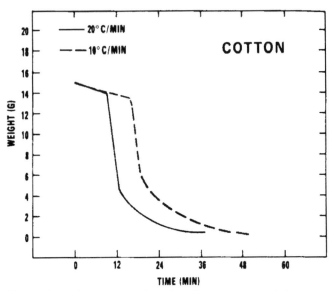

Figure 6. Weight loss of cotton as a function of time (run nos. 2-20-11 and 1-10-11).

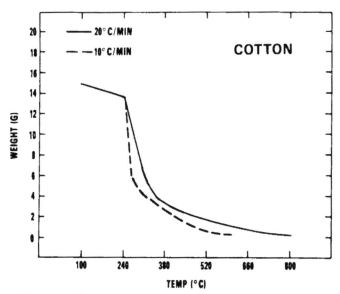

Figure 6a. Weight loss of cotton as a function of temperature.

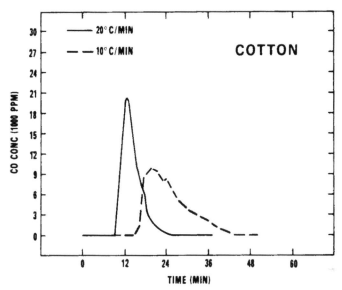

Figure 7. Carbon monoxide evolution of cotton as a function of time (run nos. 2-20-11 and 1-10-11).

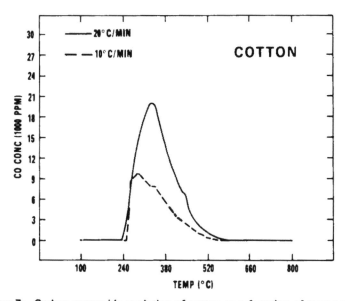

Figure 7a. Carbon monoxide evolution of cotton as a function of temperature.

311

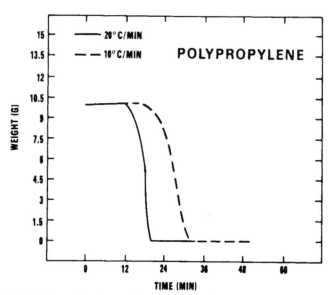

Figure 8. Weight loss of polypropylene as a function of time (run nos. 1-20-11 and 3-10-11).

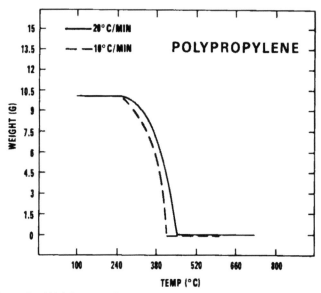

Figure 8a. Weight loss of polypropylene as a function of temperature.

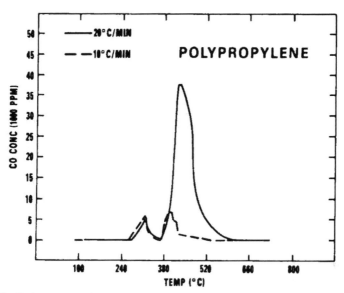

Figure 9. Carbon monoxide evolution of polypropylene as a function of temperature (run nos. 1-20-11 and 3-10-11).

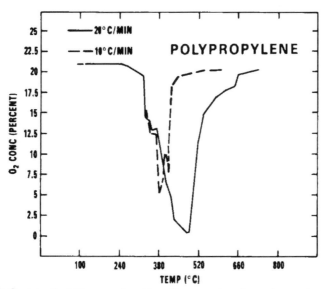

Figure 10. Oxygen depletion as a function of temperature for polypropylene (run nos. 1-20-11 and 3-10-11).

313

of oxygen at the lower heating rate was not as severe. Although it cannot be proven at this time, one hypothesis regarding the differences between these two test runs at different heating rates is that the higher heating rate caused extremely rapid (explosive) combustion, while less violent burning occurred at the lower heating rate.

Even at a specified heating rate, the behavior of the polypropylene specimens was less reproducible than the other materials tested. An example of this is shown in Figure 11, which is a plot of CO concentration versus temperature for two different runs at 20° C/min heating rate. The same test run that showed the higher CO level in Figure 9 is reproduced in Figure 11, except that it is now the smaller of the two CO plots (the range of the y-axis has been changed). The difference in the total CO produced by these two specimens, which were run in exactly the same manner, is a factor of two.

Comparison of the temperature profiles for the two different heating rates for polypropylene is presented in Figure 12. It is apparent that both specimens underwent flaming combustion; however, it is not clear from these plots why there are such discrepancies among specimens.

The mode of thermal degradation of polystyrene in these studies was unclear. It did not exhibit the severe burning behavior of polypropylene, nor did it seem to smolder like cotton. Apparently, it ignited and burned in many tests, as evidenced by small spikes in the TC-2 temperature profile. However, in other cases, CO was evolved and oxygen was consumed without clearly indicating whether the sample was burning or under-

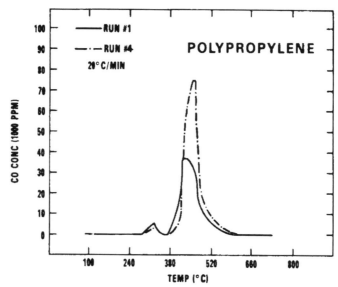

Figure 11. Carbon monoxide evolution of polypropylene as a function of temperature (run nos. 1-20-11 and 4-20-11).

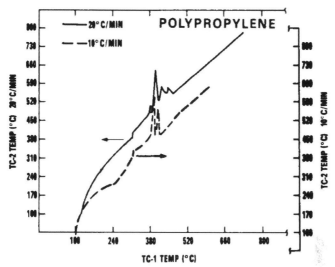

Figure 12. Temperature of a bare-wire thermocouple (TC-2) vs. temperature of the shielded controller thermocouple (TC-1) for polypropylene at two different heating rates (run nos. 1-20-11 and 3-10-11).

going oxidative pyrolysis. An example of CO plotted as a function of temperature for two runs at different heating rates is presented in Figure 13.

Flow Rate

Certain theoretical considerations must be reviewed before discussing the effect of flow rate on the evolution of CO and CO_2. The equation below describes the concentration, C, of a contaminant gas in an ideal, well-mixed flow-through system [7].

$$C = (w/b)\,[1 - e^{(-bt/a)}],$$

where

    C = concentration of contaminant gas (as a fraction)
    w = volume rate of flow of contaminant gas, L/min
    b = total rate of flow of atmosphere through the chamber, L/min
    a = volume of chamber, L
    t = time to reach "C", min.

The rate of rise of concentration, C, is dependent on the ratio of the total air flow to the volume of the chamber, b/a, while the equilibrium concentration is equal to the ratio of the contaminant gas flow to the total air flow, w/b. It is apparent from this equation that decreasing the flow rate of air, b, (e.g., from 11 L/min to 6 L/min) should cause a slower buildup of gas to a higher equilibrium concentration, all other factors being equal. The ratio of these equilibrium concentrations should be inversely proportional to the ratio of the air flow rates. However, since the generation of "contaminant" gas in these combustion experiments was obviously not constant, it is difficult to measure either the rate of rise or the equilibrium concentration. In such cases, the "total" amount of contaminant (the c-t product) should still be inversely proportional to the air flow rate.

In the experimental setup described herein, the concentrations did not follow these principles. As indicated previously, Table 4 contains the concentration-time products for CO and CO_2 for all of the runs. Table 5 presents these data for carbon monoxide from cotton only (which was the most repeatable of the various tests conducted). For any given heating rate, it is clear from the data that the total amount of CO or CO_2 is not dependent on the flow rate in accordance with the equation presented above. For example, the average values for the CO c-t for cotton at 20° C/min (Table 5) are 139, 119 and 95 thousand ppm-min for 6, 11 and 18 L/min, respectively. Likewise, at 10° C/min, the average values are 152 and 133 thousand ppm-min for 6 and 11 L/min, respectively. In the latter example, if 133,000 ppm-min is "correct" for 11 L/min, then the value for 6 L/min should be around 244,000 ppm-min (instead of 152,000 ppm-min).

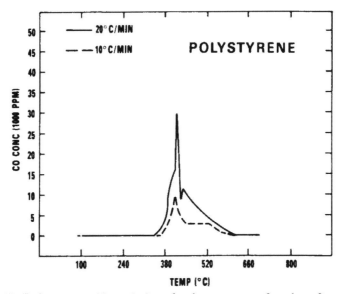

Figure 13. Carbon monoxide evolution of polystyrene as a function of temperature (run nos. 1-20-11 and 2-10-11).

Table 5. Carbon monoxide concentration-time products for cotton.

Heating Rate °C/min	Air Flow Rate L/min[a]	CO, 1000 ppm-min	
		Each Run	Average
20	6	126.0 152.1	139.1
20	11	120.6 117.8	119.2
20	18	92.2 97.8	95.0
10	6	158.4 145.5	152.0
10	11	124.0 142.7	133.4

[a]Actual air flow rates were 5.7 L/min, 10.9 L/min and 16.5 L/min, respectively, for 6, 11 and 18.

In order to resolve this question of the flow rates, several brief experiments were conducted in the test apparatus as shown in Figure 1, except that a mixing fan was placed in the furnace. Firstly, the total flow volume through the exhaust lines was verified (as had been done previously) with a dry gas meter. The air flow for the nominal 6 L/min ranged 5.4–5.7 L/min; for the nominal 11 L/min, the actual flow was 10.7–10.9 L/min; and for the nominal 18 L/min, the flow was 16.5 L/min. With everything at room temperature, pure carbon monoxide from a compressed gas cylinder was metered into the furnace space at a constant flow of 0.185 L/min over a period of 5 to 10 minutes. The appearance of the CO concentration curves at 6 L/min and 11 L/min were nearly identical, as had been observed for the actual combustion experiments and which is contrary to the theory.

After removing all possible sources of leakage in the exhaust system (namely, the quartz tube and the filtration system), and obtaining the same result as above, the furnace air space was replaced with a one gallon plastic bottle. Carbon monoxide was metered into the bottle while air was drawn through the bottle using the pump and flowmeter arrangement. The resultant CO concentrations were very nearly as calculated by the equation for 6 and 11 L/min. Redoing the experiments described above in the furnace, only with the mixing fan off, finally produced equilibrium concentrations of CO that were significantly different at the two different flow rates; however, they were both still below the theoretical values calculated for a 26 L volume.

The only reasonable explanation for the anomaly in the concentrations as functions of flow rate is that the furnace leaks. This is not too surprising since such furnaces are not designed to be air tight. However, even the presence of a small mixing fan was enough to create substantial leakage out of the furnace air space. By accident, the fan seemed to create a condition (at least for this CO flow) similar to what was observed in the actual experiments. During the combustion experiments, an unknown amount of gas was lost due to expansion and, more importantly, a different quantity of combustion products will be lost for every specimen and for every furnace.

Comparison to Literature

Alarie and Anderson both state (in private communications) that it does not matter what furnace is used. Anderson makes direct comparison of data from the University of Pittsburgh apparatus and the Arthur D. Little apparatus with no mention of the differences in furnace volumes or shapes [3]. In fact, she has even reported her 33 L, cylindrical "crucible" furnace as a 42 L "box" furnace. The 42 L volume represents a 13½ in. × 13½ in. × 14 in. "box," when in reality the furnace is 13½ in. diameter × 14 in. deep [8]. This lack of attention by these researchers to the details of the combustion device is, in the opinion of this author, a serious error.

Anderson has presented no data in the literature on carbon monoxide from her furnace. However, Alarie and Anderson presented CO data for the original combustion furnace [2,9]. In Table 6, estimated CO c-t data from Reference 2 are listed for comparison with certain CO c-t data from Table 4 (for 20° C/min and 11 L/min) for cotton, Douglas fir and polystyrene. The c-t values for the published data were estimated from their graphs, calculating time from the temperature scale (dividing by the heating rate of 20° C/min). If all of the CO evolved by any given specimen in either furnace has been measured (i.e, if there was no leakage), then the data generated in this study times the factor 11/20 (or

Table 6. Carbon monoxide concentration-time data from this study compared to Alarie and Anderson data (Reference [2]).

Material	Weight	CO c-t, 1000 ppm-min			
		This Study[a]	× 0.55[b]	Alarie[c]	(Weight)
Cotton	15 g	119.2	65.6	54.5[d]	(11.9 g)
Douglas fir	30 g	120.1	66.1	24.1	(50.0 g)
Polystyrene	6 g	93.0	51.2	4.5	(5.8 g)

[a]Average of two values each, for 20°C/min and 11 L/min.
[b]0.55 is ratio of 11 L/min (this study) divided by 20 L/min (Alarie dilution).
[c]Alarie data estimated from graphs in Reference [2]
[d]This material described as cellulose fiber.

Table 7. Comparison of CO data among this study, Alarie and Anderson
(Reference [2]), and NBS (Reference [10]).

	All values in units of ppm/mg sample charged/L air		
	This Study	Alarie	NBS
Douglas fir (flaming)	43.6[a]	7.8	83
Douglas fir (nonflaming)	167[b]	—	118
Polystyrene	169	15.4	34

[a]20°C/min heating rate.
[b]10°C min heating rate.

0.55), to calculate the effect of the dilution air, should be comparable to Alarie's data. It is apparent from the table that there are some discrepancies. The data for cotton compared favorably to that for Alarie's cellulose (65,600 ppm-min for our 15 g sample and 54,500 ppm-min for Alarie's 11.9 g sample). However, the c-t for Douglas fir in this study was more than twice that estimated from Alarie's data (66,100 ppm-min for a 30 g sample in this study compared to 24,060 ppm-min for Alarie's 50 g sample). And the CO c-t for polystyrene from this study was ten times that calculated from the Alarie plots (51,200 ppm-min for 6 g versus 4,475 ppm-min for 5.8 g, respectively).

Since it has been demonstrated in this report that cotton smolders under these test conditions, it seems reasonable that the pressure build-up in the furnace might be less than for flaming specimens. Thus, it is not unreasonable to believe that data from different laboratories for this material might not be too different (this does not mean that there was no leakage—only that the resultant CO concentrations were similar). For flaming specimens of the same size, one might expect that a smaller chamber would lose a greater quantity of combustion gases than a larger chamber (although not necessarily in proportion to their volumes). This appears to be the case for Douglas fir, where the higher CO values are associated with the larger furnace used in this study.

The above reasoning does not explain an inconsistency in Alarie and Anderson's LC_{50} values relative to their CO data. Each of the cases described above (11.9 g cotton, 50 g Douglas fir and 5.8 g polystyrene) was supposed to represent the LC_{50} for that material. If the primary lethal species in the combustion products from each of these materials is carbon monoxide, one would expect the CO c-t products from the LC_{50} quantities to be similar. Yet the calculated CO c-t data for these three specimens range from less than 5000 ppm-min to almost 55,000 ppm-min. There is no attempt to rationalize these CO data with the animal mortality.

There is another possibility, which may help clarify the anomalies. Since the creation of carbon monoxide in a combustion atmosphere may be dependent on the speed of combustion and the consumption of oxygen, one cannot assume that similar size specimens burning in different chambers will burn in the same way and produce exactly the same products. It is encouraging when there are similarities in combustion among different laboratory setups, but it seems more the rule that each combustion apparatus produces different products. Certain CO data are available from the National Bureau of Standards Interlaboratory Evaluation (ILE) of their combustion toxicity test [10]. Data for Douglas fir (flaming, F, and nonflaming, NF) and polystyrene are presented in Table 7 for comparison with data from this study and the Alarie data. The data from the ILE were reported in the form "ppm CO per mg/L"; therefore, the data from Table 6 have been converted over to ppm CO/mg/L, as shown in the equations below*. The data from this study are that from the first column of Table 6, while the Alarie data are that from the last column of the same table. It should be recalled that these data reflect CO evolution per mg of sample charged to the furnace, not accounting for weight loss. For specimens which char, this could become a significant factor in the concentration of CO per gram of material actually consumed; however, it has not been considered here since it could not be calculated from the NBS data.

Calculations in Table 7 were performed as follows:

a) This study:

$$\text{ppm/mg/L} = \frac{\text{ppm-min CO}}{\text{sample wt} \times 1000 \text{ mg/g}} \times 10.9 \text{ L/min}$$

b) Alarie data:

$$\text{ppm/mg/L} = \frac{\text{ppm-min CO}}{\text{sample wt} \times 1000 \text{ mg/g}} \times 20 \text{ L/min}$$

Neither the Douglas fir data from this study nor from Alarie and Anderson's publication agree with the flaming data on Douglas fir from the NBS test, which is not surprising. However, the data from this study (43.6 ppm/mg/L) are considerably closer to that reported by NBS (80 ppm/mg/L) than is the Alarie data (7.8 ppm/mg/L). This lends some credence to the hypothesis presented above, that the leakage from the Alarie furnace may be greater than from the furnace used in this study, although our furnace undoubtedly leaks, also. Under nonflaming condi-

---

*It should be noted that this author does not like the term "ppm/mg/L." It is used here only to reduce the number of calculations that the reader must follow. Units such as "g CO/g sample" would be much clearer and more scientific. The comparisons in Table 7, however, would be the same.

tions (10° C/min for the studies reported herein), comparison of Douglas fir data suggests that a similar total quantity of CO may have been evolved from this material under both the NBS and this laboratory's Pittsburgh test conditions (110 and 167 ppm/mg/L, respectively). No comparable data for nonflaming are available from either Alarie or Anderson.

The polystyrene CO data in Table 7 do not lend themselves to interpretation. The CO data from Alarie is lower than that reported in the ILE (15.4 and 34 ppm/mg/L, respectively), while that from the present study is much higher (169 ppm/mg/L).

CONCLUSIONS AND RECOMMENDATIONS

Since the University of Pittsburgh Toxicity Test method is undergoing consideration as a standard procedure for regulatory purposes, it is incumbent upon the developers of the method to make the apparatus and procedures as reproducible and as reliable as possible. The present study has revealed a number of possible problems in the combustion model which need to be resolved before this procedure should be adopted as a standard test method.

The problems encountered in this study, along with recommendations for further study, may be summarized as follows:

1) The combustion of materials was variable, even to the extreme of flaming vs. nonflaming, with different heating rates. It should be determined, for example, at what point between 20° C/min and 10° C/min heating rates wood changes from flaming to nonflaming combustion. Since selection of the 20° C/min "standard" rate for such a furnace was apparently arbitrary, this choice of heating rate needs to be re-examined. Introduction of a pilot ignition source for flaming conditions should be considered.

2) There was an unknown loss of combustion gases from the furnace and it seems likely that this loss would be different for every furnace. The furnace size and configuration needs to be examined and consideration should be given to reducing the sample size and/or to increasing the flow rate out of the furnace to reduce the magnitude of gas loss.

3) There was a difference in temperature of the "sample" compared to the "furnace." This should be clarified, and the "true" temperature should be verified. In the opinion of the author, controlling the rate of temperature rise by a bare thermocouple in the furnace airspace should not be the recommended procedure.

4) Certain specimens produced very high concentrations of CO_2 and severe oxygen depletion in the furnace. Not only is this not conducive to free and reproducible combustion, but dilution with air may still create a CO_2 level which is higher then desirable and an O_2 concentration lower than desirable for animal toxicity studies. This needs to be examined.

5) There was a safety hazard due to exploding samples. Furthermore,

explosions do not lend themselves to creating reproducible combustion atmospheres. This may be due, in part, to the lack of piloted ignition; however, it needs to be explored further.

ACKNOWLEDGEMENT

The author is appreciative of the valuable discussions with members of the staff of the Department of Fire Technology, Southwest Research Institute, and with others in the scientific community. Grateful acknowledgement is made to Mr. Anthony J. Valys, Southwest Research Institute, for performing the experiments reported herein.

REFERENCES

1. Kaplan, H. L., Grand, A. F. and Hartzell, G. E., *Combustion Toxicology: Principles and Test Methods*, Technomic Publishing Company, Inc. (1983).
2. Alarie, Y. and Anderson, R. C., "Toxicologic and Acute Lethal Hazard of Thermal Decomposition Products of Synthetic and Natural Polymers," *Tox. and Appl. Pharm.*, Vol. 51, pp. 341–362 (1979).
3. Anderson, R. C., et al., "Study to Assess the Feasibility of Incorporating Combustion Toxicity Requirements into Building Material and Furnishing Codes of New York State," Final Report to Department of State, Arthur D. Little, Inc., Reference 88712 (May 1983).
4. Anderson, R. C., personal communication (September 29, 1983).
5. Alarie, Y., personal communications.
6. Kaplan, H. L. and Hartzell, G. E., "Modeling of Toxicological Effects of Fire Gases: I. Incapacitating Effects of Narcotic Fire Gases," *Journal of Fire Sciences*, Vol. 2, No. 4, pp. 286–305 (July/August 1984).
7. MacFarland, H. N., "Respiratory Toxicology," in *Essays in Toxicology*, W. Hayes, Jr. (ed.), Academic Press (1977).
8. Fluegge, D. H., Lindberg, Watertown, Wisconsin, personal communication (April 26, 1985).
9. Alarie, Y. and Anderson, R., "Screening Procedures for Evaluation of Toxicity of Cellular Plastics Combustion Products," Second Annual Progress Report for Products Research Committee Grant #RP 75-2-8, University of Pittsburgh (September 1978).
10. Levin, B. C., Fowell, A. J., Birky, M. M., Paabo, M., Stolte, A. and Malek, D., "Further Development of a Test Method for the Assessment of the Acute Inhalation Toxicity of Combustion Products," Natural Bureau of Standards, Washington, DC, NBSIR 82-2532 (June 1982).

Index

323

Printed and bound by CPI Group (UK) Ltd, Croydon, CR0 4YY

23/10/2024

01778237-0011